PENGUIN BOOKS

LUCY'S CHILD

Donald Johanson is the founder and president of the Institute of Human Origins in Berkeley, California. He is the author, with Maitland Edey, of *Lucy: The Beginnings of Humankind* (Penguin 1990) and *Blueprints: Solving the Mystery of Evolution.* His latest book is *Journey from the Dawn: Life with the World's First Family*, with Kevin O'Farrell.

James Shreeve is the author of *Nature: The Other Earthlings*, the companion volume to the popular American television series. He graduated from Brown University in 1973. He is a frequent contributor to the *Smithsonian* and *Discover* magazines and is currently at work on a book about modern human origins. He lives with his family near Paris.

DONALD JOHANSON AND
JAMES SHREEVE

LUCY'S CHILD

THE DISCOVERY OF A HUMAN ANCESTOR

PENGUIN BOOKS

PENGUIN BOOKS

Published by the Penguin Group
Penguin Books Ltd, 27 Wrights Lane, London W8 5TZ, England
Penguin Books USA Inc., 375 Hudson Street, New York, New York 10014, USA
Penguin Books Australia Ltd, Ringwood, Victoria, Australia
Penguin Books Canada Ltd, 10 Alcorn Avenue, Toronto, Ontario, Canada M4V 3B2
Penguin Books (NZ) Ltd, 182–190 Wairau Road, Auckland 10, New Zealand

Penguin Books Ltd, Registered Offices: Harmondsworth, Middlesex, England

First published in the USA by William Morrow and Co. Inc. 1989
Published in Great Britain by Viking 1990
Published in Penguin Books 1991
3 5 7 9 10 8 6 4 2

Printed in England by Clays Ltd, St Ives plc

To my wife, Lenora,
for
the magic she has brought into my life
—DJ

To Chris, Marah, Luke, and
Lilah, each one and together
—JS

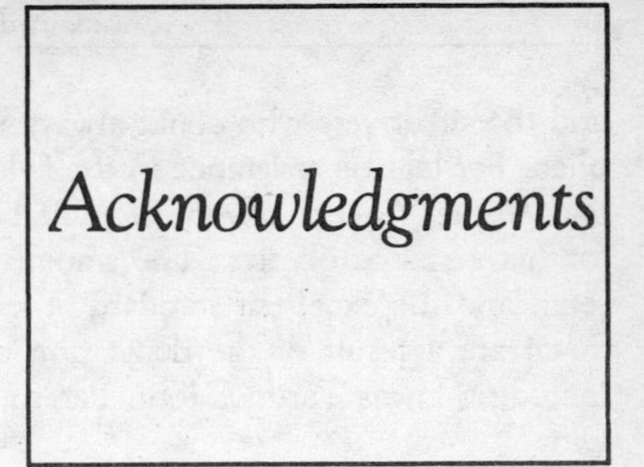

Acknowledgments

I wish to extend my sincerest appreciation first to my Tanzanian colleagues Mr. Prosper Ndessokia, Mr. Paul Manega, and Mr. Pelagi Kyauka; Dr. Simon Waane, Mr. Amini Mturi, Dr. Abel Nkini, Mrs. Jane Kessy, Mr. Omari Minazi, Mr. Harubu, Mrs. Digna Tilly, and Mr. Shadrack Kamenya in the Tanzanian Department of Antiquities; in the Tanzania National Museums, Dr. Fidelis Masao, co-leader of the Olduvai Research Project and coordinator of the archeological research, and Mzee Mrisho; Dr. Peter Schmidt (now at the University of Florida), Mr. Jonathan Karoma, Professor James Mainoya, and Professor Aswathanaratana at the University of Dar es Salaam; and Professor A. Msangi, Mrs. Lyaruu, Dr. Bitanyi, and Dr. Hirji of the Tanzania Commission for Science and Technology (formerly the National Scientific Research Council). I am deeply indebted to the above-named organizations and to the Tanzanian government generally for constant encouragement, support, and permission to undertake research at Olduvai Gorge and Laetoli.

Thanks are due to the members of the Olduvai project and visiting scientists: Lew Binford and Nancy Stone, George and June Frison, Alberto Angela, Michael Black, Hilary Wright, Jeremy Paul, Robert Drake, Robert Walter, Berhane Asfaw, Gen Suwa, William Kimbel, Carl Swisher, Robert Blumenschine, David Basiji, Stewart Patrick, Walter Hartwig, and Tim White. Thanks to the camp staff at Olduvai who kept us well fed and saw that the camp ran smoothly and to our drivers who could always be counted on to keep us supplied. For logistic assistance in the field I want to thank Sandy Evans and Annie Vincent of Abercrombie and Kent, and Hans Schneider for his aerial acrobatics. The smooth running of the Olduvai Museum and the excellent standard of lectures and guided tours of Olduvai are a result of the dedication of Peter Lauwo and his team, including Lucas Tarimo, John Pareso, Godfrey Olemoita, and Miss Dina.

I have had the delight of working with my close friend and colleague Gerald Eck in the laboratory and in Africa for nearly twenty years, and it is with sincere appreciation that I thank him for his dedication to excellence.

My appreciation also to Ian Tattersall, Fred Grine, Fidelis Masao, Prosper Ndessokia, and Tim White for their careful reading of the completed manuscript and their constructive suggestions. "Biblio Bill" Kimbel was especially helpful for his critical reading of the manuscript at various stages, his invaluable comments, and his considerable bibliographic help.

I am extremely grateful to the National Science Foundation, Gordon Getty, David H. Koch, the Ligabue Research and Study Center, and to Agip Petroleum (ENI) in Rome for their generous financial support. Gordon Hanes provided the necessary funding and expertise to upgrade the windmills and electrical system at the Olduvai Camp.

To Don Cutler and Maria Guarnaschelli for guidance in the stormy world of publishing.

Without the understanding and support of everyone at the Institute of Human Origins, it would have been impossible for this project to have been completed, many thanks to all of my colleagues there. For all of her administrative assistance and her special talent to keep track of my hectic schedule, I want to thank Larissa Smith.

Analysis and comparative study of the OH62 specimen was conducted in Cleveland and I want to especially thank Dr. C. Owen Lovejoy, Mr. Scott Simpson, and Mr. Richard Sherwood at Kent State University for stimulating discussions and assistance in the laboratory. Dr. Bruce Latimer and Mr. Lyman Jellema of the Cleveland Museum of Natural History offered critical help in our comparative studies with the Hamann-Todd Collection. Special gratitude is expressed to Jim Ohman and Picker X-Ray for the generosity and guidance in utilizing Cat-Scan equipment at their plant.

—Donald Johanson

I would like to express my thanks first to the investigators whose generous gift of time and insight made the writing of this book possible. These include Richard Alexander, Lewis Binford, Robert Blumenschine, Henry Bunn, Garniss Curtis, Robert Drake, Gerry Eck, Robert Foley, Clark Howell, Nicholas Humphrey, Bruce Latimer, Owen Lovejoy, Henry McHenry, Todd Olson, Yoel Rak, Peter Rodman, Meredith Small, Randy Susman, Elisabeth Vrba, Bob Walter, and Sherwood Washburn. There are others, equally generous, who deserve thanks, though the book eventually veered away from "the Miocene mess" that drew me to them. These include Peter Andrews, David Frayer, Leonard Greenfield, Terry Harrison, Misha Landau, Lawrence Martin, David Pilbeam, Martin Pickford, Elwyn Simons, Milford Wolpoff, and especially Vincent Sarich.

At Olduvai, I would like to express my gratitude to my hosts Prosper Ndessokia and Gerry Eck; to Berhane Asfaw and Gen Suwa for sharing their floor and passing the time (as well as the bottlecap); to Dr. Bob Walter, bane of dik-diks, and to Jim O'Connell and Kristen Hawkes for bringing some fresh perspective into camp.

In addition to the readers named above, Lew Binford, Robert Foley, John Pfeiffer, Nancy Stone, and Elisabeth Vrba all read chapters of the manuscript and provided many suggestions. I would like to especially thank Tim White for the clarity of his explanations and the rigor and thoroughness of his readings, and express my deep appreciation to my friend Bill Kimbel. These two went far beyond what could be expected of them to help bring this book into being. Victoria Pryor of Arcadia Ltd. is a writer's kind of agent, and I thank her for her part in keeping a sometimes difficult project on course. Maria Guarnaschelli at William Morrow, to my mind the best editor in New York, gave the book her intelligence, her enthusiasm, and the strength of her conviction, and I am very glad to have wandered into her domain.

My affection and gratitude to John Pfeiffer, for the inspiration he exudes and the example he sets as a writer and a friend. Finally, my loving thanks to my wife, Chris, who listened, read, edited, reshuffled, brainstormed, and then patiently listened some more. So selfless and constant was her contribution that it was sometimes all too easy to take it for granted.

—JAMES SHREEVE

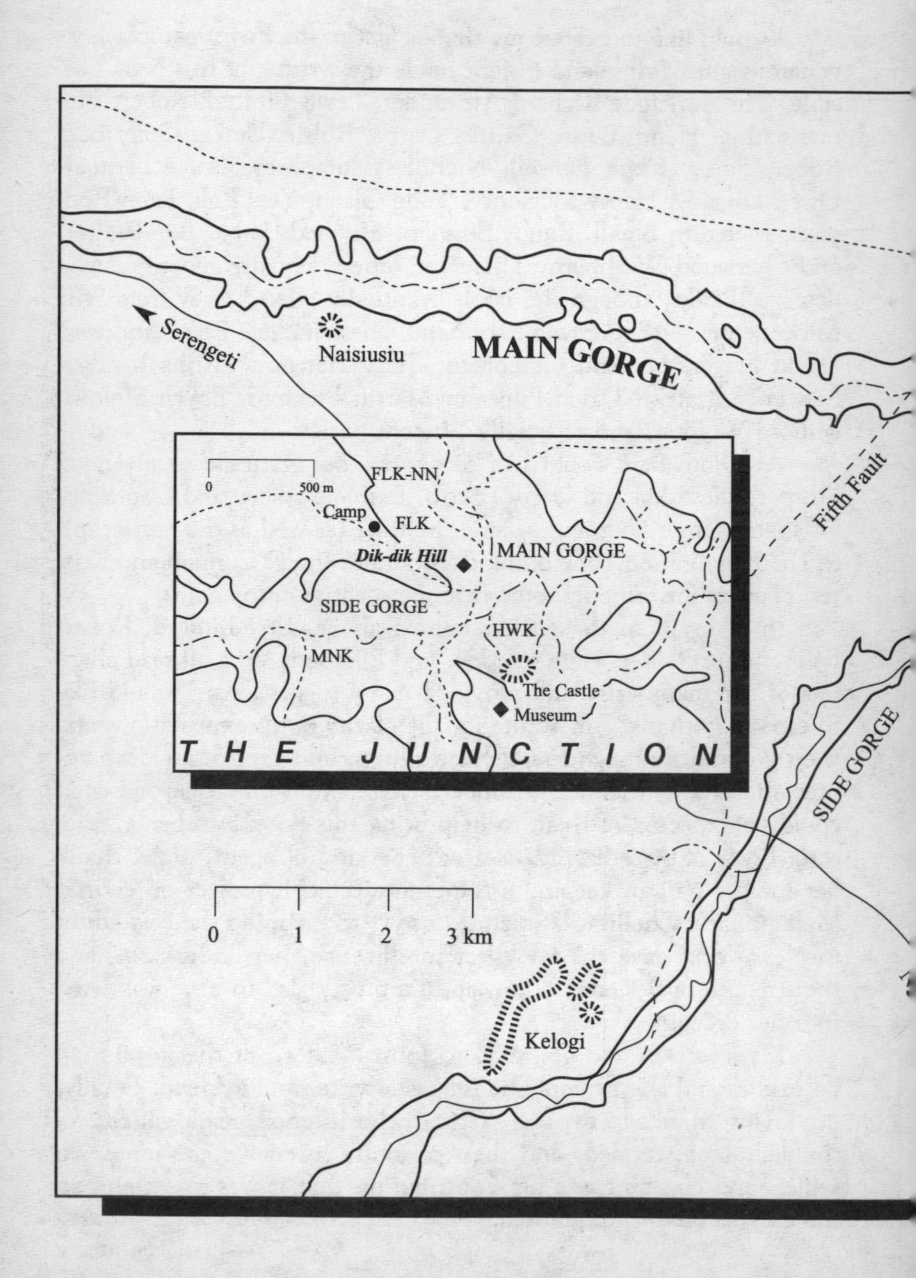

Serengeti
Naisiusiu
MAIN GORGE
Fifth Fault
0
500 m
FLK-NN
Camp
FLK
Dik-dik Hill
MAIN GORGE
SIDE GORGE
HWK
MNK
The Castle
Museum
THE JUNCTION
SIDE GORGE
0
1
2
3 km
Kelogi

OLDUVAI GORGE

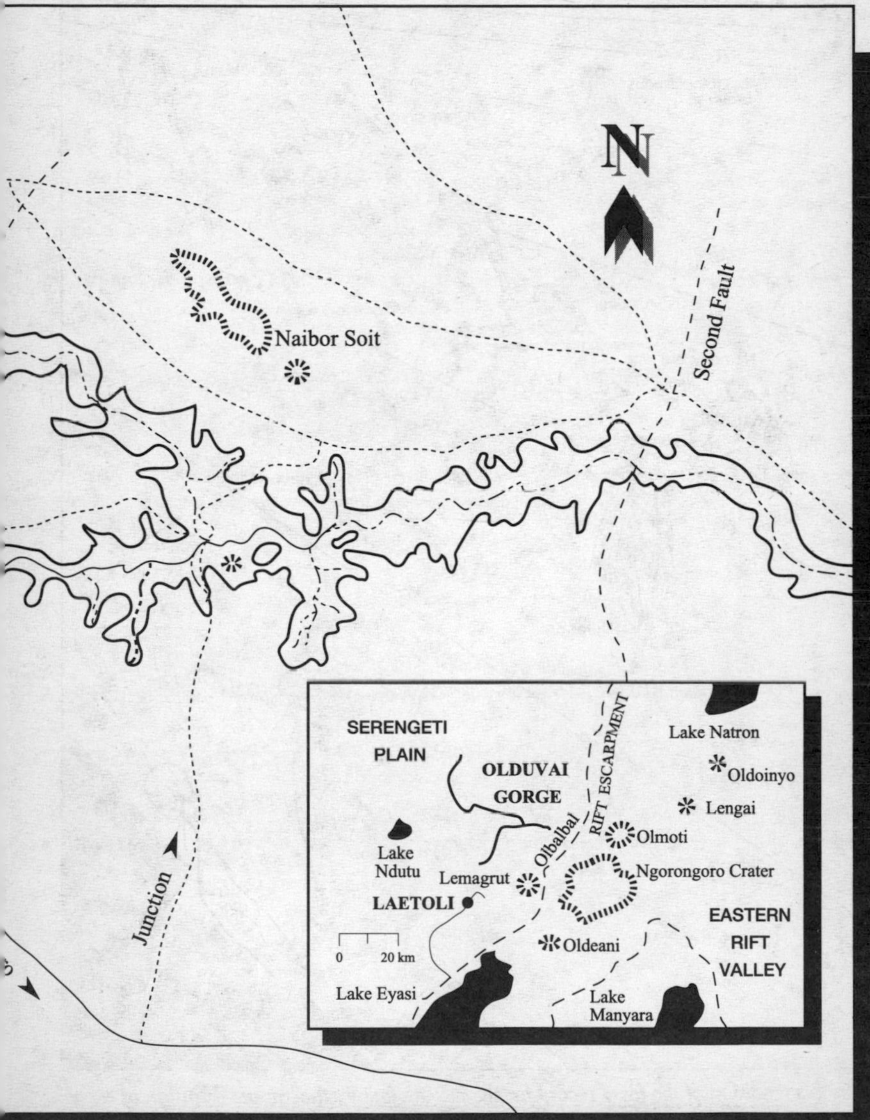

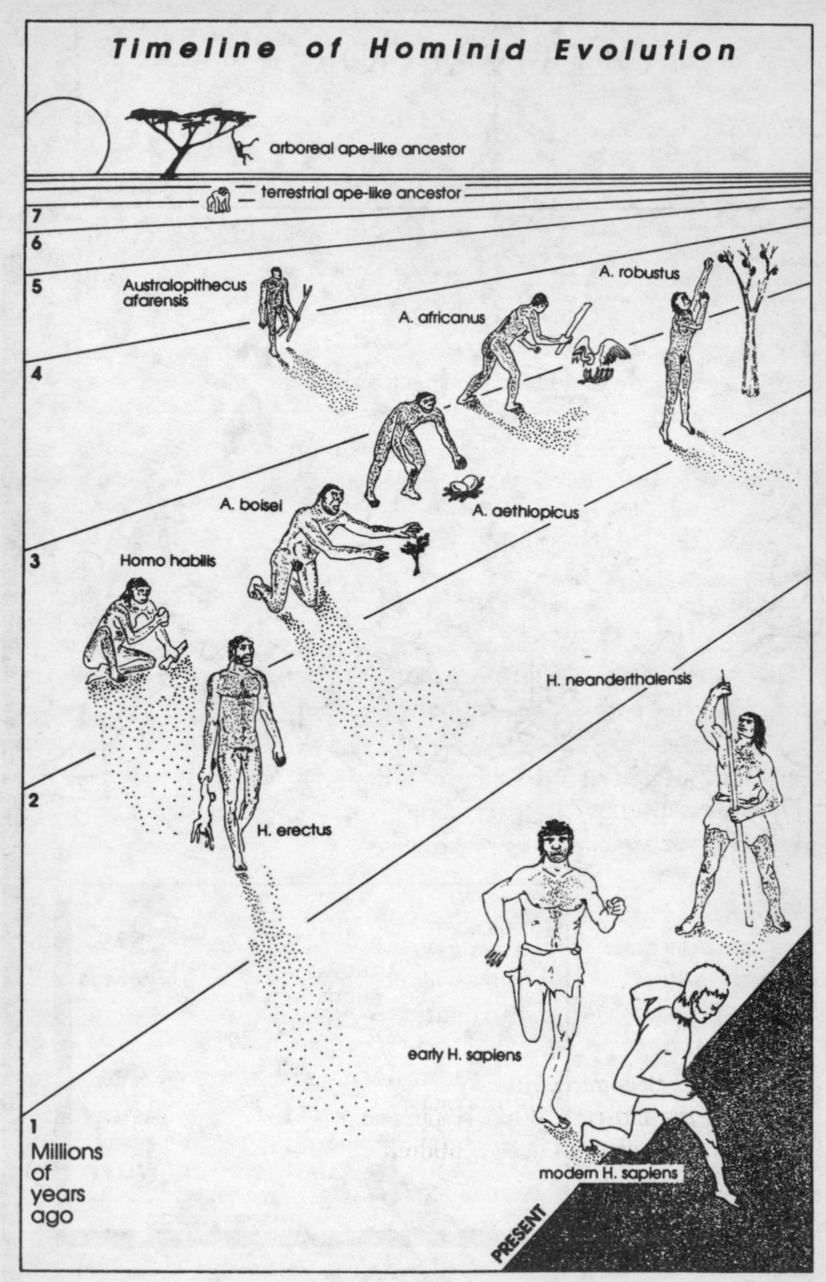

In this rendition of the fossil record for the human family, the known duration of each species is represented by the shadows cast by the figures. The exact relationship between extinct hominid species is a matter of interpretation, and their placement here is not meant to suggest a particular phylogeny. DESIGN AND GRAPHIC BY DOUGLAS BECKNER. AFTER RICHARD HAY, *GEOLOGY OF OLDUVAI GORGE*, UC PRESS, 1976.

Contents

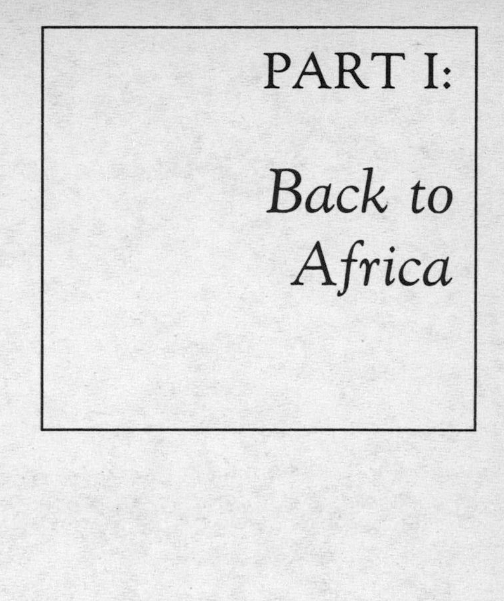

PART I:

Back to Africa

CHAPTER ONE: *Cold Reflections*

Once more on my adventure brave and new.
—Robert Browning

AS SOON AS MY EYES opened I reached for my watch on the bed table. I knew already it was close to 3:30 A.M. Over months of restless sleep I'd developed an ability to sense what time it was the moment I found myself awake, as if every minute through the long night had its own familiar scent that I could instantly recognize. Guessing the time had become a sort of grim pleasure, a little bit of consolation for being awake in the first place. It didn't matter whether I was sleeping at home in Berkeley, in a hotel room in Ames, Iowa, or in Addis Ababa, Ethiopia. It made no difference when I went to bed, or what seemed to have brought me out of sleep in the first place—a shout outside the window, a nervous dream, a knotted back muscle. I nearly always guessed within ten minutes of the clock.

I wonder if all insomniacs share this useless expertise.

On this night I was in a room on the second floor of a hotel in Dar es Salaam, Tanzania. In a few days I would join up with colleagues for a field expedition I was leading to Olduvai Gorge, some four-hundred miles to the northwest. I am a paleoanthropologist. I have spent the last twenty years trying to understand the origins of our earliest ancestors—how they evolved from apelike predecessors, what they looked like, how they moved and what they ate, and ultimately, what novelties of behavior led them to become the most powerful beings the world has ever seen. To answer these questions, I look at the evidence they have left behind—primarily fossilized bones. Ironically, most of what we think about these ancient ancestors comes from fossils uncovered only during the past half a century. Olduvai Gorge had played a big part in that new influx of knowledge—so big that anyone about to lead an expedition there should have been able to excuse himself for not being able to sleep. Especially if he had hardly worked in the field for nearly a decade. Especially if it seemed his whole reputation and future in science had come down to what he might find when he got there.

In the dim glow from the window I could just read the watch face: 3:28. Not bad. With a piercing brightness I suddenly remembered the dream that I had been having just before I awoke. I was standing on the edge of a deep, cool lake bordered with palm trees. The surface of the lake was mottled with a shifting swirl of color, as if someone had poured rich pigments into the water and let the currents take them away. People—friends of mine, mostly, fellow anthropologists—emerged from behind the palms, speaking in hushed tones about something of obvious import. In pairs and small groups they approached the lake, and then one by one they dove from the bank, turning somersaults as they fell. Each disappeared into the water without a splash. Suddenly I saw myself running along the bank, waving and shouting. I wanted to dive too, but something kept pulling me back from the edge. I was all alone, but around me the air was filled with a bustling murmur, a sound like that of a hundred whispered conversations taking place all at once. It began to grow cold. I hugged my shoulders and shivered.

Now awake, it struck me that I was in fact shaking with cold. Beneath the window an air-conditioning unit gave forth a pulsating hum, blowing frigid air toward me across the empty twin bed between. I got up to examine the unit, dragging the bedspread along with me. The air conditioner was housed in a long, vented metal box painted the same dingy beige as the walls of the room. Underneath a panel on the top there was a knob. I tried to turn the machine off with my fingers, but the knob refused to budge. The rushing of cold air from the vent below seemed to shoot out and grab hold of my ankles. I hopped up and down on the bare linoleum floor, wrapping the bedspread tighter around me.

On field expeditions I always carry with me an all-purpose tool, one of those ingenious devices that you can transform into a knife, a pair of wire cutters, or fifteen other implements by flipping open the right sequence of hinges. At this moment I desperately needed the thing to become a pair of pliers. I found the tool in the bottom of my duffel bag, and with a little fiddling it assumed its plier persona. The air conditioner hummed and snorted, louder than ever. I grabbed hold of the knob with the pliers and turned it full to the left. Nothing happened. I gave it a full twist to the right. Nothing still.

I yanked off the covers on the unused twin bed and spread them out on my own bed, together with a surprisingly thick blanket I found in the closet. Crawling in, I pulled the covers up to my chin and stared up at the ceiling. Only then did it strike me how strange it was to be lying awake in a hotel room on the tropical African coast, huddled up with cold.

People say I am a lucky man. From 1973 to 1977 I co-led a series of expeditions to Hadar, part of a remote and desolate region of Ethiopia called the Afar Triangle, two hundred miles northeast of Addis Ababa. We returned from these expeditions with fossil specimens beyond our wildest hopes. In 1973, when I was barely out of graduate school, I found a humanlike knee joint that proved beyond doubt that our ancestors walked erect close to three and a half million years ago—long before they developed the big brains that had once been thought to be the hallmark of humanity. The next year, while a colleague and I were looking for fossils on the slope of a gully, just before lunchtime, I looked down and noticed a piece of an elbow joint lying on the ground. A few feet away I spotted the back of a skull. A thighbone turned up next, then some vertebrae, a part of a pelvis, ribs . . . suddenly it dawned on us that we might be standing among the remains of a single humanlike individual, a skeleton more complete and millions of years older than anything like it yet known. That was a lucky prize indeed.

In three weeks' time the expedition team had collected and pieced together nearly half the skeleton, with no duplication of parts that would suggest there was more than one individual. Without question it was a hominid—a member of the family of erect-walking primates to which we, *Homo sapiens,* belong and of which we are the sole surviving representatives. Because of its small size and pelvic structure we could tell that the specimen was female. We named her

"Lucy," after the Beatles song "Lucy in the Sky with Diamonds" that was playing in camp on the day she was lifted from the obscurity of that little gully. It was largely on account of Lucy that our work became known to an audience beyond the scientific community. But my luck did not run out with that find. The next year we returned to Ethiopia and uncovered an even more exciting treasure—a collection of more than two hundred hominid bones from a single site, representing at least thirteen separate individuals, perhaps all struck down in a single event three million years ago. These fossils came to be known as "The First Family." With the consent of the Ethiopian government, we borrowed the whole Hadar collection and took it back to the Cleveland Museum of Natural History, where I was curator, to begin the long process of analysis.

Yes, I have been lucky. The Hadar finds in fact turned out to be a gold mine for all of us who study human origins. Paleoanthropology is a science that is forever rooted in chance. But the luck that nourishes its progress has little to do with the kind that wins a hand of blackjack or finds a parking space at midmorning in downtown San Francisco. Consider what it means to find a hominid skeleton. When an animal—any animal—dies on the savanna, the scent of its decay immediately attracts scavengers of all kinds. They tear apart the hide, devour the soft tissue, crush the bones in their jaws. Very little remains uneaten. The few bone fragments and teeth left behind will be bleached white by the sun, and in three or four years disintegrate into dust. They will leave no trace of their existence.

A very few animal carcasses, however, suffer a different fate. Their scattered remains sink slowly into river silt or lake mud, where they lie hidden from the sun. If the soil is too acidic, the bones will disintegrate after all. With the right chemical balance, however, the process of fossilization begins. It is a slow metamorphosis, still little understood by chemists, of bone tissue into stone. Meanwhile, the centuries pass, piling hundreds of feet of earth between the fossils and the open sky. Of course, not all the fossilized bones will be hominids. Even when two or three hominid species shared the savanna habitat two million years ago, they accounted for a tiny percentage of the mammals found there. And even among the hominid remains that are turned to stone, all but a handful remain deep in the earth, buried forever.

Millions of years after their burial, and only in certain parts of the world, violent shifts of the earth's crust deliver a few bones near to the surface again. Ten thousand more years of rain and wind claw at the ground, and then one day a spring storm flushes the last bit of dirt away: A fossil lies exposed again on a rocky slope. Even then, the chances are fantastically remote that a trained fossil hunter will come snooping around in just the right place. More likely, the bones will just be washed away, or splintered into fragments by years of fluctuating moisture and temperature, or trampled to dust by generations of wildebeest moving across the savanna.

But let's say the fossil hunter does arrive in time. Let's say too that he happens to take precisely the right pattern of footsteps as he edges across the slope. If the sun is slanting down at just the right angle to catch the luster of bone against the muted dust, and if at that moment he should look down and see a fragment of his ancestry gleaming at his feet, then yes, I would say that he is a very lucky person indeed. *Every* fossil find is miraculous. Luck, of course, is not the only advantage one needs to bring into the field; most discoveries have been made by skilled collectors who through years of experience have developed the ability to spot fossils virtually invisible to the untrained eye. Nevertheless, any field-worker who has succeeded against such overwhelming odds acknowledges in his heart the debt he owes to luck—if such an uncanny conspiracy of volcano, earthquake, rain, wind, and time can be distilled into such a paltry word.

In all, the Hadar site yielded up some 6,000 fossils, of which 250 proved to be hominid—jawbones, teeth, skull fragments, limb bones, toes, fingers. In a matter of a few seasons we had multiplied by many times the record available from a crucial and dimly understood period in human evolution, three to three and a half million years ago. Eventually each fossil would play its part in the unfolding of a whole new story of the origin of the human lineage. My colleagues and I proposed a new ancestral species for mankind. Together with one of those colleagues—Tim White of the University of California at Berkeley—I used this new species to construct a new human family tree. Because of Lucy and the other Hadar finds, I would be asked to lecture across the country, appear on television, be host to a popular nature program. With science writer Maitland Edey, I would write a book about the Hadar discoveries and what they tell us of the human career. The book was a best seller, translated into Russian, Dutch, Japanese, and half a dozen other languages. Later I moved to Berkeley to establish a new institution for the study of early man. More interviews would follow, more requests

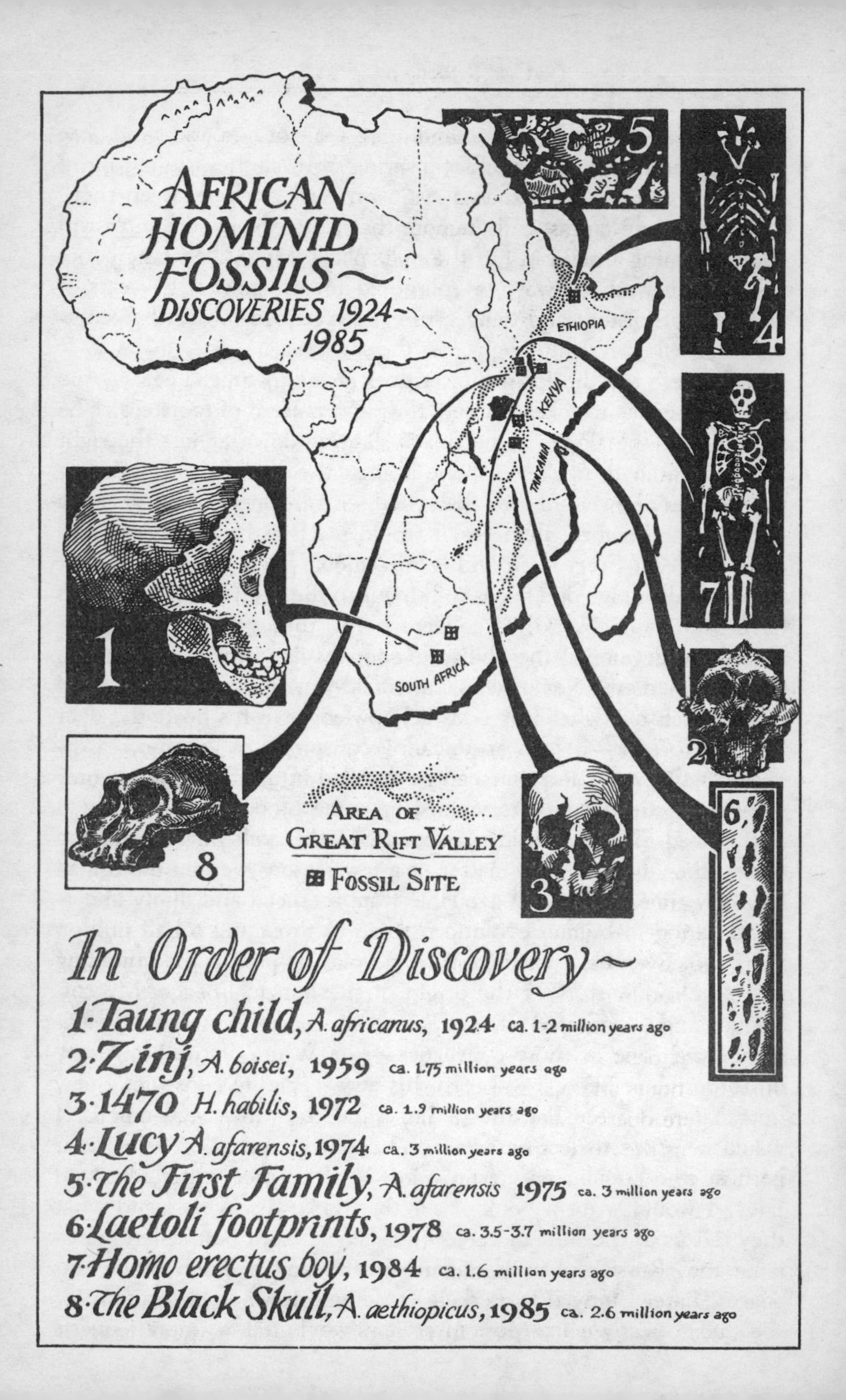

AFRICAN HOMINID FOSSILS~
DISCOVERIES 1924–1985
ETHIOPIA
KENYA
TANZANIA
SOUTH AFRICA
1
2
3
4
5
6
7
8
AREA OF GREAT RIFT VALLEY
FOSSIL SITE
In Order of Discovery~
1·Taung child, A. africanus, 1924 ca. 1-2 million years ago
2·Zinj, A. boisei, 1959 ca. 1.75 million years ago
3·1470 H. habilis, 1972 ca. 1.9 million years ago
4·Lucy A. afarensis, 1974 ca. 3 million years ago
5·The First Family, A. afarensis 1975 ca. 3 million years ago
6·Laetoli footprints, 1978 ca. 3.5-3.7 million years ago
7·Homo erectus boy, 1984 ca. 1.6 million years ago
8·The Black Skull, A. aethiopicus, 1985 ca. 2.6 million years ago

for lectures and appearances. I was introduced to royalty, given awards, and came to be regarded as a public spokesman for American anthropology.

But the celebrity Lucy brought me had a bitter side as well. Along the way, I would also be called a prima donna, a slick operator, a publicity hound. I lost friends, including some of my closest colleagues in the field, whose interpretations of humanity's origins were thrown into serious doubt by Lucy and her Hadar companions. My hopes for a return to Ethiopia to complete the work we had started there would be frustrated again and again. Other field expeditions yielded nothing. Slowly I would come to doubt my famous luck. Eventually I would find myself alone in a hotel room in Dar es Salaam, shivering the night through and wondering whether I was finished as a practicing scientist.

I was pulled from my thoughts by the air conditioner. Suddenly its throaty hum modulated upward to a hysterical whining pitch, shaking its side panels with a sound like the rattling of chains. I was beginning to begrudge it a will of its own, a single-mindedness of purpose that commanded respect. The temperature continued to fall. I pictured myself frozen where I lay, discovered in the morning by a maid. I imagined her standing in the subtropical hallway wearing a light cotton dress with short sleeves: She knocks and enters, letting out a grasp as the arctic air hits her solar plexus. She marvels at the icicles hanging from the lampshades, the walls caked with layers of hoar. Then her eyes reach the bed, and there she finds a man dressed in his socks, underwear and turtleneck, encased in a clear block of ice.

Alarmed by my own imagining, I considered getting up to alert the night manager. But then I remembered a conversation at the hotel front desk, soon after I'd arrived. I'd been traveling for over thirty-six hours and was in need of a shower.

"There doesn't seem to be any hot water in my room," I told the clerk at the desk.

"We're sorry, sir, but the water heater is temporarily out of order," she said.

"How long has it been broken?"

"Just over a year," she said.

I thanked her and went back to my room, reminding myself on the way to take such inconveniences in stride. Westerners have an overblown sense of authority over their environment. We fully *expect* hot water to pour from every faucet, effortlessly and on demand—as if heated water running through a system of metal pipes were part of the natural order, like gravity or sunshine. Each time I come to Africa, I rediscover how much we live at the mercy of our own contrivances. It's not so much that things fail to work; it's that they don't always work according to our expectations of how, when, and in what fashion they *should* work. Once, for example, I found that the phone lines in the city where I was staying were down and it was impossible to call across the street. But the next minute the hotel switchboard put me through on a call to my mother in California, and she sounded as clear as if she had been speaking from the next room.

As if to confirm my thoughts, the air conditioner suddenly gave out a series of bone-jarring thumps and turned itself off, leaving behind a self-satisfied silence, as if it were confident that there was enough cold air stockpiled in the corners to last till morning. But after a few minutes I could sense the warmth begin to creep palpably back into the room, first along the walls and then edging toward the bed in luscious, swirling drafts. I got out of bed, still wrapped in the bedspread, and walked over to the window. Dawn had just touched the tops of the buildings, a soft, pearly light sketching in the outlines of the streetscape—the rambling market buildings and in the near distance the bulky facade of an old office building. Across the road I could just read the CAL TEX sign above a gas station. Not a soul moved. Dar es Salaam in Arabic means "haven of peace," but this was the first time I'd seen it living up to the name. By day it is a bustling metropolis, the sidewalks crowded with people and merchants' stands, the streets laden with overstuffed buses riding low on their suspensions, leaning precariously into every turn.

I worried about what the day would bring. When I'd arrived the morning before, I had been met at the airport by Dr. Fidelis Masao, an archeologist and director of the National Museum of Tanzania. Masao had been instrumental in arranging the Tanzanian side of the expedition to Olduvai. A handsome, athletic man of inexhaustible good spirits, Masao nevertheless had a worried look on his face when he shook my hand.

"We've had a little trouble securing permits for all the visitors to Olduvai," he explained. "The permit to survey at Laetoli has yet to be approved as well."

"Who's not been approved to go yet?" I asked. I tried to sound casual, but was aware of an abrupt sinking sensation, giddily unpleasant.

"Binford and Frison and their wives. Perhaps some others."

This was distressing news. Lewis Binford and George Frison were archeologists we were relying on to bring a fresh perspective to the famous Olduvai excavations of early human-occupation sites. A problem with access to Laetoli, twenty miles to the south, could be even worse. The fossil-bearing deposits at Olduvai end abruptly in a bedrock of basalt 2 million years old; Laetoli's sediments extend backward in time almost twice as far. Already they had yielded a fine collection of hominid teeth and jaws, at 3.5 million years among the oldest indisputable hominids in the world. And it was at Laetoli ten years before that a member of Mary Leakey's team had made the sensational discovery of a trail of footprints, left behind in wet volcanic ash by some ancestors to the human species, 3.5 million years ago. Olduvai was to be our base of operations, but secretly I was pinning my hopes on what we might find at Laetoli.

"How serious a problem is this?" I asked, throwing my suitcase in the trunk of Masao's car. He shrugged.

"I think not very serious. Most everyone is excited about getting things under way."

As we drove into Dar, Masao brought me up to date on the status of preparations, going over arrangements and acquainting me with officials we would need to meet with the next day to secure final permissions for the field. What he said was reassuring. He had scheduled a meeting for me with the Principal Secretary for the Ministry of Community Development Culture, Youth and Sports, responsible for overseeing the joint expedition. Masao believed I would find her wholeheartedly behind the joint Tanzanian-American expedition. Indeed, the permit delays seemed to be simply a matter of bureaucratic tangles. But I still could not get over that sinking feeling—it was all simply too familiar.

In 1982, I had flown into Addis Ababa, Ethiopia, overjoyed to be on the brink, finally, of a return to Hadar. The expedition had been long in coming. Following our successes in the mid-1970s, it was predictable that a certain amount of competition would develop among anthropological teams trying to gain a foothold in the vast, unexplored deposits of the Afar Triangle. Anthropologists tend to be a competitive lot under the best of circumstances. In this case, at stake was nothing less than a share of the richest fossil deposits ever discovered. I was still surprised, however, when rumors began to circulate in Addis that I and my colleagues were incompetent, or that we were intentionally trying to shut others out. There was no truth in any of this. Nevertheless, the government stopped issuing permits to work in the Afar. All we could do was be patient and wait.

By 1982 the situation appeared to be brightening. In the previous fall a team of anthropologists led by Berkeley archeologist Desmond Clark had been granted permission to explore a region in the Afar called the Middle Awash. The team unearthed precious finds in the field, including some four-million-year-old hominid fragments. But the expedition's greatest significance lay in its long-range hopes. Among the members were some young Ethiopians, including Berhane Asfaw, a graduate student in anthropology at Berkeley. He had come to the Afar to accumulate field experience that would serve him well later on. Rather than simply play host to foreign investigators, Ethiopians like Berhane would eventually earn their doctorates and assume responsibility themselves for coordinating the development of their country's spectacular paleoanthropological resources. Meanwhile, through the effort of Desmond and other Berkeley scientists like Tim White, Clark Howell, and myself, a new laboratory and storage facility for collected material had been built at the National Museum—another step toward establishing a nationally based science of anthropology in Ethiopia.

The success of the Middle Awash expedition in 1981 had kindled great hopes for the 1982 trip back to the Afar. But when I had flown into Addis and met Tim at the airport, he had looked nervous and tired.

"Don't rush to unpack," he told me. "We've got trouble, and it looks bad. The government has declared a moratorium on all foreign research into prehistory."

"A ban? Now? How could this happen?"

"That's what we've spent the last week and a half trying to figure out," Tim said. "In a nutshell, until they've got formal guidelines written up for coordinating research, they don't want foreigners in here. To tell you the truth, I can't really argue with that. It's just been too haphazard in the past. Somebody gets a permit, excavates, finds a fossil. Then off it goes on loan, to be studied in some other country, written up in a foreign journal by authors with foreign names.

These are Ethiopian national treasures—you can't blame the people here for wanting to take control of them."

"But that's precisely what you've been trying to accomplish by establishing the center at the National Museum," I said. "Good storage facilities, controlled access to the collection . . ."

"The problem is not what we're doing or not doing. The problem is what some of our so-called colleagues *say* we are doing. There are rumors going around that the new lab is for *our* exclusive use. I've even heard we've begun training Ethiopian students only recently, to make things look good. That really burns me up. Desmond has been training local students and academics in his expeditions for decades."

"So what do we do now?"

"I don't think there's anything we can do to salvage this season's work," Tim said.

He was right. After meeting with officials the next day, it was clear that we would get no closer to Hadar that season. They could give us no guarantee for field research the following year either, or at least until the new regulations could be written and put into effect.

In a profession crowded with *if*'s, it pays to develop a thick-skinned optimism and be prepared for anything. We agreed that we should assemble all the equipment for a field season in 1983, to be ready in case the ban was lifted. Once those details were taken care of I flew home, hoping once again for a chance to be back the following year. But there was not to be any expedition to the Afar in 1983, nor in 1984, nor the year after.

It was easy, under the circumstances, to feel frustrated by all that had happened. A few years earlier, Tim and I had proposed one of the most controversial theories of human evolution in the last fifty years. In a nutshell, we suggested that Lucy's species, which we called *Australopithecus afarensis,* was the common ancestor of all subsequent hominids, including *Homo.* If we were right, then everyone who believed that *Homo* was much older than Lucy's three million years was wrong—including Mary Leakey and her son Richard, who upon the death of his father Louis had assumed his place as the most famous living paleoanthropologist. The evidence Tim and I needed to further test our theory could be found only in the rich deposits of the Great Rift Valley, a Y-shaped scar in the earth's crust that runs north through Tanzania, Kenya, and Ethiopia. This gigantic continental seam has been pulling apart for millions of years, to the accompaniment of volcanoes and earthquakes. The violent geological history of the Rift Valley, delivering long-buried strata to the surface, has been an enormous boon to fossil hunters, including myself.

But Tim and I were affected by rifting of another sort. Ten years earlier, we could regard Mary and Richard Leakey as our friends and collaborators. Since then both of us had feuded openly with the Leakeys, who controlled the principal research sites in Kenya and Tanzania. The likelihood that we would ever be welcomed to work in those countries was extremely remote—and now Ethiopia was indefinitely closed as well. Thus, through a strange mating of ancient geology and modern politics, we were effectively shut out of East Africa, with no access to the fossils that could prove us right or wrong.

As it was, I had little enough time to spend brooding over lost opportunities. In 1981 I had left my position at the Cleveland Museum of Natural History to establish the Institute of Human Origins in Berkeley. For years it had been my dream to create a research institution devoted solely to the study of prehistory—a think tank that would bring together scholars from all over the world, sponsor field expeditions, and serve in training scientists and technicians from the United States and from the countries where the fossils were to be found. Berkeley was a perfect location for such a resource. Though the Institute would not be affiliated with the university, it could work in collaboration with members of the anthropology, earth sciences, and paleontology faculties—people like Clark Howell, Tim White, Desmond Clark, Garniss Curtis, and the others who made Berkeley the leading center for paleoanthropology in the world.

With the help of some interested supporters, IHO began operations in a small basement room just off the Berkeley campus. All we had were some surplus desks, chairs, bookcases, and a rented IBM typewriter. It was a modest beginning, but given time I imagined the Institute maturing into a real force in the field. In my mind's eye I saw shelves upon shelves of fossils and casts accessible to all who would come and take advantage of them. I imagined the most up-to-date radiometric dating labs, where geologists could verify the ages of the latest fossil discoveries. I envisioned casting and molding labs, journal collections, film libraries, publishing enterprises, conferences, symposia—and most of all I imagined IHO field expeditions in operation in East Africa and beyond.

Dreams like that get a hold on you and won't let go. They also take a lot of money to keep alive, much less to mature. We needed two hundred thousand dollars just to start, and another half a million to staff and supply the Institute for the first two years. Far more would have to be raised for the endowment we would eventually require. On our side we had people like Gordon Getty, Diana Holt, Hubert Hudson, David Koch, Arnold and Joan Travis—all of them passionately dedicated supporters who dove headfirst into the challenge of keeping the fledgling Institute on its feet. But I knew that its ultimate survival would depend on my success at raising more funds, more interest, more support. I devoted myself to appearances and lectures, channeling all the honoraria I received back into the Institute. I was learning a lesson from the Louis Leakey School of Anthropology: To keep yourself in the field 10 percent of the time, keep yourself on the stump the other 90 percent.

As a student I had listened enraptured to the famous Dr. Leakey's lectures, and I knew what made them such a success. In spite of how much he privately griped about the demands made upon him, Leakey did not lecture simply to raise money. He lectured because he loved to talk about his work. The huge auditoriums crowded with the curious worked on him like a drug; he inhaled the public's fascination with human beginnings and boomed it back at them a hundredfold. I too counted on the eagerness and expectation of the audiences to inspire me, and they rarely let me down. Wherever I went, people turned out to hear the man who had found Lucy, the man who could tell them more about their origins. I lectured in great halls in New York, Chicago, and Los Angeles. I lectured in college classrooms in Palo Alto and Boston, in high school auditoriums in Salem, Massachusetts, and Bozeman, Montana. The reception was no less energetic overseas. Desmond Clark and I once lectured together in a little town west of Venice called Rovigo, where they packed a thousand people into a tiny amphitheater and turned away hundreds more.

The public interest in human origins is itself a phenomenon to be reckoned with. Ever since Darwin proposed that we were not always as we are now, people have been consumed with an urge to know what we were before. Thomas Huxley, Darwin's champion on the lecture circuit, talked to houses packed with scientists and laypeople alike. When in 1925 an obscure anatomist named Raymond Dart announced that he had identified the skull of a man-ape found in a South African lime quarry, the headlines ran in banners around the world. Louis Leakey never spoke of his discoveries in a lecture hall that wasn't mobbed with people squatting in the aisles and practically hanging from the light fixtures. His son Richard is no less popular today.

Without diminishing the contributions of these investigators, I think that Lucy appeals to the public imagination in a special sense. Part of her attraction is her great age. Lucy burst into the twentieth century straight from the remotest beginnings of our past. Before she was found, paleontologists had only a handful of older specimens presumed to be hominid. A piece of an arm bone believed to be 4 million years old. A shard of lower jaw dated at 5.5 million years. A single molar, half a million years older still.

All of these fossils are weatherbeaten, anonymous fragments. Lucy is not a piece of bone. She is a *skeleton*, a work of organic architecture much like ourselves. We found 40 percent of her on that hillside in Hadar. But because we had pieces from both sides of her body, and because the two sides of a skeleton are mirror images, we were later able to accurately reconstruct over 70 percent of her. There is immense scientific value in a specimen so complete, but Lucy's substantiality excites more than our intellects. Enough of her has been recovered from the past for our imaginations to fix upon and fill in the rest, the way we can gather the gist of a conversation just by listening to stolen snatches of it. Only in this case, it is a conversation taking place three million years in the past.

There is something else that draws us to Lucy, beyond her age and completeness. I know this is a provocative thing to say, but I think Lucy's popularity has a lot to do with her sex. A piece of jaw or a fractured skullcap has no gender, and if we endow them with one, it will tend automatically to be male. In our society male is still the undifferentiated standard, the "generic brand." Lucy becomes all the more recognizable for being female. Whether or not one believes, as I do, that her species was a direct ancestor of humankind, her sex grants her the rudiments of a human identity. And I think it also taps a very deep well in our collective imaginations. In an elusive but powerful sense she represents the Mother, Gaea, Isis—or whatever history has called the fertility that lingers at the beginnings of our conciousness.

As a scientist, I am treading dangerous ground by speaking of Lucy in "mythic" terms. Lucy's real significance, of course, is not to

Ato Woldesembet Abomisa, storekeeper of the National Museum of Ethiopia, with skeleton of Lucy
DONALD JOHANSON

be found in symbol; it rests in whatever empirical evidence she provides for the understanding of the process of evolution, specifically the evolutionary origins of the human species. Nevertheless, it is important to acknowledge that for many people, the search for those beginnings excites the same regions of the mind once served by the sustaining myths of religion. This places a heavy responsibility on those of us engaged in prehistorical research. When Louis Leakey toured the United States twenty years ago, he told of ancient men at Olduvai building stone dwellings, slaughtering herds of giant buffalo, or tripping up game with rounded stones tied together with thongs. Leakey was speaking the truth as well as he knew it at the time, but much of what he said has since been disputed. In the last decade especially, paleoanthropology has hardened itself as a science, developing ever-more-precise methods and greatly expanding the mass of evidence that lies at its roots.

In the process, a rigorous skepticism has put into shadow the liberally subjective methods that once epitomized the field. The more we know about our origins, the more we must question assumptions long regarded as fact. This critical caution has sometimes been misinterpreted by the so-called "creation scientists" to be a sign of weakening of conviction among paleoanthropologists in the conclusions of evolution science. Just the opposite is the case. It is a sign that paleoanthropology is at last coming of age.

I had an obligation to my field to get that point across to the public—an obligation that seemed on a collision course with the public's expectations of what I had to say. As I lectured around the country, I began to be bothered by the feeling that my words were being interpreted as having some special authority. I looked into the faces in the audience—businessmen stiff in their chairs, undergraduates scribbling wildly in notebooks, elderly people smiling encouragement—and I could see myself mirrored in their eyes as the representative of my own ineffable truth, as if I'd been to the Mount and brought back a crisp, spanking-new version of the Beginning, this one etched in logic and the solemn surety of science. I could see the hunger for that truth in their faces, and hear it repeated in the scratching of the pens in the notebooks. And I could hear its echo in the questions people would ask after the lecture. What makes us human? When did mankind develop a moral sense? What is the future of our species?

Rather than inspire me with self-importance, the questions only filled me with a nagging sense that something was going wrong. I am an anthropologist, not a priest. Donald Johanson's opinions on the future of mankind have no more intrinsic worth than those of any other reasonably informed individual. More to the point, even my views on human origins—including my firmly held interpretation that Lucy's species was the direct ancestor to the human line—are necessarily tentative. I would be the first to reject that hypothesis if new evidence proved it false. But it didn't seem to matter how many qualifiers I put in my talks, how many *maybe*'s and *in my opinion*'s I laced into the text. I stared out at the people and could feel, like a weight, their ready concurrence: *Johanson said it; that's the way it is.*

There was another, more fretful reason for that sense of things-gone-amiss. I knew of course that the people in that auditorium had come to hear me because of what I had already accomplished in the Afar. But how long could I use the past for my platform? Ever since the opening of the Institute in 1981 I had been spending more and more time keeping up with its demands—fund-raising, lecturing, hiring staff, administering. It was not simply the lack of a place to do field research. Less time than I liked to think was going toward the nitty-gritty stuff of the science, the analysis of specimens and the writing of interpretive papers. It was all I could do to keep up with reading the journals in the field. From what I read and through discussions with colleagues, I was well aware of the advances that were taking place. But what part would I play in them? Serving the role of a public spokesman for paleoanthropology suited me well—but did that mean that I could not be a paleoanthropologist at the same time? And if I was no longer a working paleoanthropologist, what right did I have to speak for those who were?

Meanwhile, the Institute was growing toward my original vision. In 1983 we moved into larger quarters, sharing a building, quite amiably, with the Church Divinity School of the Pacific, an Episcopalian seminary. A highly sophisticated laboratory for dating fossil deposits, headed by Garniss Curtis and Bob Drake, was set up in the basement. Upstairs a staff was busily working up ideas, writing grants, drafting papers. Already we had hosted conferences and participated in several expeditions overseas. The paleoanthropological lab downstairs was nearly in place. The chemicals used in casting and molding lay in boxes on the floor. In one corner, a cast of Lucy was spread out on a table among the newly installed fixtures and apparatus. But the blue, felt-topped workbenches were clean, the cabinets uncluttered. With no route back to Africa, it all seemed a magnificent creation waiting for a purpose, like a thoroughbred horse with no track to run upon.

The promise of an end to the frustration came unexpectedly. One day in the spring of 1985 I was dictating some correspondence when Gerry Eck walked into my office. Eck is an old friend and colleague, an expert on fossil monkeys who also happens to be brilliant at managing field expeditions. My first expeditions to Africa had been in the Omo of Ethiopia, and it was "Field Marshal" Eck who had kept the Omo teams fed, supplied, and moving forward on four wheels—a monumental task, given the isolation of the region and the extremities of the climate. Since 1974 he had been teaching at the University of Washington in Seattle.

Eck is never wholly calm—he always seems charged up and ready to do battle with some nagging problem that only he fully appreciates. On that day he was almost twitching with excitement.

"I've got something for you," he said. "It might be nothing, or it might be something very big."

"Let me guess. Richard Leakey's decided to let us have five minutes alone with the Kenya National Museum specimen of our choice."

"This is serious. What do you know about Prosper Ndessokia?"

"He's a Tanzanian student working in the Berkeley paleontology department. He was with Glynn Isaac on the Lake Natron expedition in 1982. Nice guy. Working on baboon skulls."

"Which is why I went over to see him in the first place. But not why I'm here."

"So what's the point?"

"Ndessokia is a paleontologist in the Tanzanian Antiquities Department. According to him, an application from us for a permit to work in Tanzania would be a good idea. It might even get us into Olduvai Gorge."

"Forget it, Gerry. Mary Leakey would never give us permission to work at Olduvai," I said. Mary had retired the previous year and moved to Nairobi to finish writing up a monograph on her most recent excavations at the Gorge. She would still have an interest in what went on there. And though Richard Leakey's authority over research sites ended at the Kenya border, I felt he might have something to say about who worked the Gorge as well.

"That's just what I told Prosper. He was shocked that I thought Mary had any authority over issuing permits for research in Tanzania. That's the Tanzanians' prerogative alone. And that applies to Olduvai, Laetoli, the whole shebang."

"Wait a minute. Are you saying we could work at Olduvai Gorge without Mary's consent?" I asked.

I must admit that at first the notion seemed oddly unsettling. For anyone with an interest in human origins, the name *Olduvai Gorge* is indelibly intertwined with the name *Leakey*. Mary and her husband, Louis, had put Olduvai at the very center of the paleontological map in East Africa with a series of brilliant discoveries beginning in the 1950s. While Louis had subsequently gone off to spend more of his time elsewhere, Mary stayed on at the Gorge, painstakingly excavating the archeological remains. Her monograph on the stone-tool industries at Olduvai is a classic in the field. Laetoli was less well known, but the footprints Mary had found there in 1978 ranked with the greatest discoveries ever made.

"I'm just passing on what Prosper told me," Gerry replied. "According to him, the Tanzanian National Scientific Research Council would be annoyed at the suggestion that we had to get clearance from an Englishwoman or a Kenyan national in order to work their sites, no matter what their last names happen to be. Not to take anything away from Mary Leakey, but she's not going to do more work in Tanzania, living in retirement in Nairobi."

"I think we'd better have a talk with Ndessokia," I said. "Soon."

Temperamentally, Prosper Ndessokia turned out to be as cool and unflappable as Gerry is nervous. His countenance suggested that no matter how chaotic life seemed to be on the outside, matters were slowly moving toward balance and resolution underneath. When we met with him the next day, Prosper bore out everything Gerry had told me, and more. The Tanzanian Department of Antiquities already had formulated ambitious, long-range plans for developing its country's rich prehistorical sites, including the training of Tanzanian scholars. The National Museum had decided to establish a natural-history museum in the regional capital of Arusha, a hundred miles east of Olduvai. But while the future seemed promising, the present was stalled for lack of expertise and funding. Save for the small staff who operated the little museum on the rim of the Gorge, the permanent camp at Olduvai had been vacant since Mary's retirement.

With Prosper's encouragement we applied to the National Scientific Research Council, or Utafiti, for a permit to conduct fieldwork in collaboration with Tanzanian scientists and students for a period of three years. On a morning in early May, my colleague Bill Kimbel greeted me at the door of the Institute, his dark, bearded face lit up by an irrepressible grin. He handed me a telegram from Mrs. A. E. Lyaruu, an official of the Utafiti. APPROVAL HAS BEEN GRANTED TO ENABLE YOU AND YOUR ASSOCIATES UNDERTAKE STUDIES FOR YOUR PROPOSED PROJECT RENEWED PALEOANTHROPOLOGICAL RESEARCH AT OLDUVAI GORGE. . . .

When I read those words my heart soared. Good news had been a long time coming.

"We're *in,* Bill," I yelled, "we're in the Gorge!"

Kimbel frowned. By nature he is cautious and methodical. Over the decade we've spent working together Bill has become accustomed to dampening my outbursts; we joke that he keeps a pitcher of cold water in his desk drawer for just such emergencies.

"Hold on to the hysterics," he answered. "We're not anywhere yet. That piece of paper means nothing until we've got the funding to put us in the field. God knows how we're going to work it out in time for a field season this year. . . ."

"So what's your prognosis, Dr. Killjoy?"

"Get Eck on the phone. I'll call Tim and tell him to get his butt over here. We've got some thinking to do."

Over the next couple of days, Tim, Bill, Prosper, Gerry, and I—along with geologist Bob Drake in the dating lab downstairs—took stock of what had to be done. It was patently clear to us that there was little hope of a full-scale expedition that year. Just to prepare the grant proposals for funding would require a review of all that had already been accomplished at Olduvai Gorge, which was after all the most thoroughly scrutinized piece of anthropological real estate in the world. Next, we would have to mount a convincing argument for why the Gorge deserved more study, not only for the possibilities it held for further hominid discoveries, but for the understanding of its geology, paleontology, archeology, and ancient environments. None of this would be difficult, we thought, but it would be time-consuming. The deadlines for most grant agencies had already passed.

In the end, we decided our best strategy was to take things one at a time. First, conduct a brief survey trip to Olduvai, financed by the Institute, to assess what Olduvai really had to offer. On the way, firm up relationships with the Tanzanians, get to know their priorities. Finally, draft proposals to finance full-scale operations in the summer of 1986. We had decided that Tim, Gerry, Prosper, Bob, and I would make the initial journey to Tanzania in July; Bill would stay behind to keep the Institute running.

Before leaving there was one more matter to attend to. Since our first discussions with Prosper Ndessokia, I had not communicated to Mary Leakey our hopes to renew research at Olduvai Gorge. For reasons I will go into later, my relationship with Mary had collapsed precipitously in 1978, and we had not spoken to each other since. I knew, of course, that she would not be exactly overjoyed to hear that I was going to be leading a research expedition at the site that for so long had been her workplace and her home.

Until we were sure we were going to Olduvai, I saw no reason to upset her by parading our hopes. If I had learned anything about the business of anthropology over the years, it was to resist the temptation to count unhatched chickens. More important, since Olduvai was a Tanzanian site, it was up to the Tanzanians to contact Mary. For us to tacitly seek someone else's blessing on the proposed research seemed to me at the very least inappropriate, and possibly insulting to our hosts.

In June, soon after we had received the telegram from the Utafiti, I sent a letter off to Mary letting her know we had been issued a permit to conduct research in northern Tanzania, specifically at Olduvai Gorge and Laetoli. I also promised to keep her informed of any developments. She wrote me back in July, expressing considerable alarm at our intentions and suggesting that I was "ignorant of the normal scientific procedures under such circumstances."

I have tremendous respect for all that Mary has accomplished at

Dar es Salaam, capital of Tanzania DONALD JOHANSON

Olduvai. But if I had violated any procedures at all, they were the unwritten customs of the "old boy" brand of anthropology that had held sway in East Africa since colonial times: The Leakeys had long held a claim to develop the Gorge, so they were entitled to some influence in deciding who would move in after they were gone. In my mind, however, it was time to stop thinking of the Gorge as the Leakeys' living room. It deserved a future as well as a past, and clearly that future was in the hands of the Tanzanians themselves. I even hoped that once she was over the initial shock, Mary would realize it was better that Olduvai should continue to contribute to paleoanthropology rather than lie fallow while potentially valuable fossils washed away in the seasonal rains.

In early August of 1985 we flew to Dar es Salaam and got down to serious discussions with officials in the Department of Antiquities, the Utafiti, and the University of Dar es Salaam. Our mutual strategies would focus on three areas—research, site management, and education. Well-known areas like Olduvai and Laetoli would be the immediate focus of research, opening up the famous Leakey sites to fresh approaches. Later, Olduvai's extensive facilities could also serve as a launching pad to explore new sites to the south. Plans would be developed to protect Olduvai's fossil deposits, both from the hooves of Maasai cattle and from the hands of the tourists who might unwittingly remove irreplaceable material from the site. A program of teaching and field training would be outlined for Tanzanian students, beginning with Prosper Ndessokia and his counterpart in geology,

Paul Manega. Recommendations would be made on how best to deal with the famous footprints of Laetoli. Several years before, Mary had reburied the footprints in an effort to protect them from the elements. It was a reasonable stopgap measure, but buried footprints were of little use to the science. More important, they were not immune from damage just because they were underground—water could seep down through the soil; acacia roots could burrow down and split them apart. How could the footprints be permanently preserved? Could they be excavated and transported to the proposed new museum in Arusha?

All of this would take years of effort and considerable funding. To start, we would have to convince granting agencies that the primary sites of Olduvai and Laetoli held as much promise as we thought they might. Olduvai does not contain vast, unexplored horizons, as do Hadar or the Omo. Generations of fossil hunters have left their footprints on its slopes; hundreds of pairs of eyes have peered into every little curve and recess in its fabric. Geologists have walked out, labeled, analyzed, and dated its deposits. Archeologists have excavated key sites of hominid activity. Paleontologists have labored over the meaning of its faunal remains, scrapped all their conclusions and worked them through again.

We knew that some observers might see Olduvai as a poor prospect, rather like a bargain bin in a department store that had already been picked through for months by keen-eyed shoppers. But in our calculations, those hundreds of eyes from the past would give scientists studying the Gorge a tremendous advantage over those working newer sites. To anthropologists, Olduvai connotes far more than a site for research. It has come to mean an incomparable body of knowledge, a well of information unmatched in its richness and depth. If there were new clues yet to be squeezed from its horizons, they would lock readily into place amid a complex, highly ordered system of data—rather like new pieces in a jigsaw puzzle that already has its borders complete.

Ten days were all we could afford in the field in 1985, just enough time to sweep quickly through the Gorge and take stock of what it had left to offer. As we climbed over its slopes and gullies, we weren't fooling ourselves about our prospects of making a sensational find. But we could be assured that whatever scraps of wisdom it still had to divulge would be doubly illuminated by all that had gone before.

What we found in those brief few days was more than enough to convince us that the famous old site was far from exhausted. The rains the spring before had been particularly heavy, and everywhere fossil bones and ancient stone tools were eroding from the slopes. In the bottom of one gully Gerry found a partial elephant jaw. A few days later, Bob Drake came upon a complete fossil cranium of a rhino. As a test of one particularly rich slope, four of us cleared every bone fragment and stone tool from a five-by-five-meter square. In thirty minutes of hard work we had recovered over seven hundred bits of bone, eighty-eight tooth fragments, one snake vertebra, and over two-hundred pieces of stone that we could see had been fashioned into crude tools by hominids millions of years in the past. We hoped that the evidence for the sheer density of fossil accumulations would help impress granting agencies of the need to carefully survey the rest of the Gorge the next year. When we returned, the cleaned patch of slope would also show us how much new material had been flushed out by a year of continued erosion. A side trip to Laetoli also revealed rapid erosion of the fossil beds. Numerous mammalian bones, three and a half million years old, dotted the landscape.

Having accomplished as much as we could, we packed up and headed home. Through the fall we worked on grant proposals to the National Science Foundation and the National Geographic Society, outlining an ambitious plan to survey the Gorge over three years and explore Laetoli's bush-covered deposits as well. We invited Lewis Binford and George Frison to join us for a few weeks of the field season. I was particularly interested in what Binford would have to say. A sharp-witted, often caustic iconoclast, Binford had already published a volume highly critical of the Leakeys' interpretations of Olduvai's ancient past, and I knew he would welcome a chance to study the sites firsthand. Meanwhile Bob Walter, a geologist who had helped establish the stratigraphy and dating of the Hadar deposits ten years before, offered ideas for expanding the absolute dating framework of Olduvai's deposits. For the first time in years, I felt charged up and on top of something new, heading a team of professionals with a promising site and an ambitious mission.

Then one day in May, Bill Kimbel came into my office with bad news. Our final proposal to the National Science Foundation had been turned down. The National Geographic Society, which had funded the Leakeys at Olduvai for a quarter of a century, had rejected another proposal a few days before. We called Tim over from his lab on the Berkeley campus, and the three of us grimly shuffled through the stack of grant reviews, trying to understand where we had come up short.

In academic science, grants are awarded or turned down by the "peer review system." Copies of proposals are sent out by the granting agency to other professionals in the field who have the expertise to judge their relative merit. After writing up a critique of the grant, each reviewer summarizes his or her evaluations in a rating, ranging from excellent to poor. After all the reviews are in, the ratings are averaged and the grant is given a rank in relation to all the others being considered. Awards are then made according to how much money is available to each field in that particular granting period—perhaps all of those rated "very good" will be funded, but if money is tight, only those with consistently excellent ratings may be supported. Peer review is by no means a perfect system, but it is one way to ensure that in a highly competitive environment, the best projects have the best chances of being funded.

The anonymous reviews of our proposal were decidedly mixed—not surprising, perhaps, in a field distinguished by some bitter sparring between competing camps.

"With the access, facilities and support of Tanzanian government and researchers, Olduvai should be the big news of the next decade," wrote one reviewer, rating the proposal excellent.

"Not systematic," wrote another, suggesting that the NSF would be paying for scientists simply "to clamber around looking for hominid fossils with absolutely no guarantees." He or she checked the box marked "fair."

"For whatever reasons," noted a third, "neither Olduvai nor Laetoli have been fully tapped for their treasures . . . It is thus rather exciting to see some promise of 'more to come' from a group of experts with extensive field experience, cross-disciplinary ties, and the apparently active support of Tanzanian authorities."

But the next one I turned up in the pile believed just the opposite: "Given the number of years of research at the site, it is not too likely that the team will find much more hominid material." That reviewer also gave us only a fair rating.

"This one says it's 'a huge gamble,' " Bill said, handing me a white sheet. "But it still rates the project 'very good.' What do you make of that?"

"Get a load of this one," Tim grumbled, " 'The dice are heavily loaded against the discovery of significant quantities of good new hominid material.' What is this, a crap shoot?"

" '. . . somewhat improbable that marked new quantities of new hominid specimens will be located,' " I read. "Are you beginning to detect a theme here?"

"Yeah," Tim answered, "Nobody believes Olduvai has anything left to give. Or at least that's what they *say*."

"What do you mean?" Bill asked. "What aren't they saying?"

"I'll paraphrase it for you: 'If you guys think you can go out and raid the Leakeys' garden, go ahead and find your own dough,' " Tim replied.

"I don't hear that," Bill said.

"Then listen to this one," said Tim. " 'Omission of any mention of the Kenya National Museum is so striking that it might be interpreted as deliberate and could suggest that there are political undercurrents in the proposal.' Since when is a proposal for research in Tanzania obligated to mention a museum in a neighboring country, just because its director is Richard Leakey? Forget the undercurrents—the politics in that review are pretty obvious."

Then Tim turned to me with one of his wide, lopsided grins.

"So how about it, Johanson? Don't you think it's time to put on one of your pinstripe suits? There must be some people out there who still think good paleoanthropology is worth supporting."

"We don't have much time," I said. Nobody had an answer to that.

Later that week I was scheduled to participate in a "Celebrity Chef" contest in San Francisco to benefit the March of Dimes. I was looking forward to the event as a break from Institute anxieties. The judges for the contest included some very big names in cuisine—Jeremiah Tower from the Santa Fe Bar and Grill, Wolfgang Puck from Spago's in L.A., and Craig Claiborne, among others. I'm a decent amateur cook, good enough to know that any razzle-dazzle concoctions would not impress chefs of that caliber. Better to keep things simple. I chose a recipe that Gerry Eck had made for me once, chicken thighs prepared in an orange cream sauce.

For a *sous-chef,* I'd drawn Gordon Getty's wife, Ann. We set up our booth at the formal evening gathering and put on aprons. Gordon appointed himself our official cheerleader, and the three of us unwrapped the ingredients and set them out on a board. I browned the chicken in some foaming butter, slapped a cover on, and let it cook for half an hour. Meanwhile, friends drifted by and kibitized. When the thighs were almost done I put them aside, grated some orange and lemon rind into the pan, splashed it with some wine and sherry, and stirred in a half pint of heavy cream. When the sauce was well thickened I eased the chicken thighs back in and let them bathe until they were done. Late in the evening, the prizes were announced by Beverly Sills. The winner for entrée—Chicken Thighs in Orange Cream Sauce. We collected an ice bucket for our efforts.

It was all just a lot of fun, and toward a good cause. But in the back of my mind I couldn't stop worrying about Olduvai. Maybe some of that worry showed on my face.

"Bill Kimbel told me about the grant problems," Gordon said to me when we finally sat down.

"They don't think there's anything left to find," I said.

"So what do you think?"

"Maybe there is, maybe not. But it's wrong to think of the Gorge as some overfarmed piece of bottomland with nothing left to give. Those slopes are steep. One heavy rain could change the whole complexion of the place, I'm sure of it."

"It seems to me that it's worth the gamble," said Gordon. "I'd like to help get you there, this season."

Gordon was a generous benefactor, but he was also a friend. If I were to approach him for funding, it should be under different circumstances.

"You've done a lot for the Institute already," I said, and meant it.

"Then think of me as just another fan of IHO cuisine," Gordon laughed. "I've been swept off my feet by Chicken Thighs à la Eck."

Gordon's "chicken grant" was the beginning of a hectic few days. By the end of the week we had received limited support from two other sources—David Koch and the Ligabue Research and Study Center in Italy. The expedition would not be the full-scale effort we had hoped for, but at least we would be in the field for a couple of months.

On July 12, 1986—two days before my cold encounter with the hotel air conditioner—I had flown to London to catch a KLM flight to Tanzania. We landed at Kilimanjaro International Airport just after midnight, well ahead of schedule. Inside the terminal I drank a cup of coffee and waited to embark on the last leg of the journey to Dar es Salaam. I looked out the observation window at the huge DC-

10 parked on the tarmac, its loading bays flung open, fuel trucks and baggage carts bustling around beneath the plane's enormous silver belly. The flight crew descended the gangway, looking worn out. On the far edge of the circle of light and activity, the eye of an African hare shone briefly, and I could just make out the animal's form as it skipped off down the runway and into the blackness. Finally, we reboarded the plane and took off for Dar. I watched the sun rise from thirty thousand feet.

Fidel Masao was at the airport to meet me, bringing with him the report of possible permit problems. By then I was exhausted, my anxiety caused as much by jet lag as it was by the disconcerting echo, in his news, of Tim's somber greeting at the airport in Addis, four years before.

"There's something else you should know," Masao said as we drove toward Dar. "It seems Mary Leakey has hired a lorry to drive out to Olduvai. From what I hear she's removed everything that wasn't nailed down—tables, chairs, mattresses, beds, stoves, casts, books, perhaps even the windmills and generator. It's hard to know exactly what's going on."

"Wonderful. What's Eck done about it?" I asked. Gerry was scheduled to have arrived in Dar ahead of the rest of us to purchase equipment and supplies. Then he was supposed to move on to Arusha, where we would all rendezvous on the seventeenth.

"What do you think Gerry would do?" Masao grinned. "He's spent the last two weeks putting furniture together, whenever he wasn't searching the bazaars for pots and pans. When he finally had everything packed and ready to go in the Antiquities Department lorry, it blew a cylinder liner. That took two more days to replace."

"Good old Eck. Maybe we'll have edible food out at the camp, maybe not. But he'll make damn sure we have a place to sit and eat it."

Masao dropped me off at my hotel, with a promise to come by after breakfast the next day. I had been traveling for thirty-six hours straight. I discovered the lack of hot water, and after my fruitless inquiry at the desk, settled for a cold sponge bath over the sink. I went to bed with my mind in two worlds. In a few days' time—if all went well—we would be descending the Ngorongoro Crater onto the famous Serengeti Plain, Olduvai Gorge a dark furrow in the near distance. Over and over again my thoughts rehearsed the moment when the camp gate would appear around a bend in the road, the sun drenching the dusty grass in the yard, people moving about, establishing camp, *working* again, at last. I imagined what it would be like if, while "clambering around looking for hominid fossils," we actually found something of value. But I still could not turn off the part of me fretting over details of business back home—phone calls at the Institute I'd left unreturned, budgets and proposals to complete, lectures to prepare. And then there were the permit problems to face in the morning. Through the wall came voices and muted music from a portable tape player. I fell asleep to the sound of a reedy saxophone playing "New York, New York" over the indifferent rhythm of an electric band.

As things turned out, Olduvai Gorge did have another secret to reveal that summer. True to its African nature, the discovery would frustrate all our expectations, cause us grief and confusion, and still leave us utterly enchanted. But I am getting ahead of my story. Like every hominid fossil, the new discovery takes on meaning only in the context of all the discoveries that have come before it. In 1859, in a quiet little cadence to *The Origin of Species,* Charles Darwin hinted that his theory of evolution by natural selection might throw light on the true beginnings of mankind. Ever since, people have been worrying the earth for traces of our ancient ancestors. But the past is very deep, and the clues to its meanings are few and fantastically subtle.

Dr. Giancarlo Ligabue examines fossil-bearing sediments at Olduvai Gorge. The Ligabue Study and Research Center has collaborated with the Olduvai Project since 1986.

CHAPTER TWO:
The Antiquity of Man

That then this Beginning WAS, is a matter of faith, and so infallible. WHEN it was, is matter of REASON, and therefore various and perplex'd.

—John Donne

"DESCENDED FROM THE APES!" EXCLAIMED the bishop's wife to her husband. "My dear, let us hope that it is not true, but if it is let us pray that it will not become generally known."

Her prayers notwithstanding, knowledge of Darwin's theories spread like a fire through all strata of Victorian society; the first printing of *The Origin of Species* sold out in one day. The idea of evolution itself—the development of new species out of older, more primitive forms—was not original with Darwin, and not everyone in late nineteenth-century England found it as unpalatable as the bishop's wife. There was even something decidedly attractive about the idea. Evolution could be thought of as the biological equivalent of the Victorian faith in progress—a gradual, ladderlike progression of experiments, each one slightly "more fit" than the last, leading ultimately and ineluctably to life's supreme refinement, man himself. For a society bursting with the self-congratulatory fervor of a colonial power and the potency of an industrial revolution, it took very little effort to prop this biological premise under a social construct and extend the ladder just a bit to justify the white man's inherent superiority over the "lesser" peoples of the earth.

Thus evolution was a tolerable concept, even a jolly good idea. What rankled was the mechanism Darwin had proposed as the force driving it forward. According to his theory of natural selection—Darwin's *real* heresy—the physical and biological environment that a species inhabits determines which of its members will survive and reproduce the most offspring. The traits that adapted the surviving individuals to their environment will be more prevalent in the next generation, and the species will evolve accordingly. Thus all species—including our own—are not the end product of some process of internal refinement; they are simply the material result of environmental pressures acting upon random variations in the traits represented in each generation of individuals. This apparent lack of design, divine or otherwise, was distasteful even to many of Darwin's supporters, and his detractors found it completely indigestable. "A law of higgeldy-piggeldy," sneered one eminent scientist, while another cast it as "a moral outrage." No wonder: Through the lenses of Darwinian selection, man was no more the apotheosis of creation than was the mosquito who sucked his blood, or the microbes that lived on his excrement.

Darwinism's strongest opponents were in fact not bishops and their wives, but other scientists. Leading the attack in England was the famous anatomist Richard Owen. Owen attempted to squelch the notion of man's descent from the apes on the basis of explicit anatomical evidence. While he acknowledged some similarities between *Homo sapiens* and the great apes (chimpanzees, gorillas, and orangutans), Owen labored to impress upon his fellow scientists that there were irreconcilable differences as well. He focused in particular on differences in structure that would not be influenced by environmental pressures, and therefore could be counted on to remain the same through unending generations. For example, Owen offered, consider the prominent brow ridges featured on the skulls of gorillas. Since no muscles attach to these ridges, it is hard to envision what behavioral role they might play in the animal's relationship to its environment, and thus they must have remained unchanged through the gorilla's ancestry. Human skulls, on the other hand, do not have prominent brow ridges, and their *absence* would be a human trait no matter how far back in time one wishes to press the mater. Ergo, gorillas and humans can never have been related.

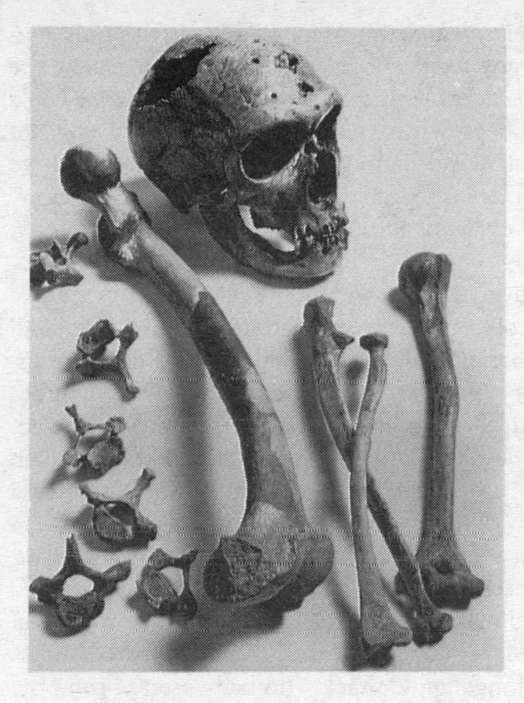

Neandertal skeleton found near La Chapelle-aux-Saints, France, in 1908. The curvatures in the limb bones, and other deformities, led scientists to believe that Neandertal walked with a stooped, bent-kneed gait. Decades later, it was shown that these were in fact the remains of an old individual crippled with arthritis.
JOHN READER/SCIENCE PHOTO LIBRARY

In 1856—still three years before the publication of Darwin's great work—some strange bones were found by workmen in a quarry cave in the Neander Valley near Düsseldorf, Germany. The bones appeared to be human, but weren't like those of any normal human alive: The leg bones were bowed, the limbs exceptionally robust, with pronounced muscle attachments suggesting enormous strength. The flattened skull of the specimen seemed to have housed a brain just as large as that of a modern human, but Richard Owen notwithstanding, the skull's eye sockets were shaded above by very prominent brow ridges. Whatever Neandertal Man was—and the attempts to answer that question gave vent to some of the most imaginative thinking in the history of science—it could be ignored neither by the evolutionists nor by their opponents.

Like it or not, Darwinian evolution was at least a workable scientific hypothesis. It could make predictions that could then be tested with evidence. Darwin assumed that proof of his theories, including testimony to our common ancestry with the apes, would be borne out from the accumulating fossil record. But two decades after he published *The Descent of Man* in 1871, the evidence for that relationship was still meager. Neandertal Man was about the only material on which to base any judgment. Naturally, the bones drew the intense scrutiny of scientists on both sides. If the pro-Darwinists could prove that they indeed represented some earlier step in the ladder leading to man, then mankind had evolved from a more primitive condition. If the anti-evolutionists could somehow explain them away, then the paleontological evidence for man's more primitive past would be reduced to zero.

From the quarry cave near Düsseldorf the odd bones first made their way to the laboratory of a German anatomist named Hermann Schaaffhausen. After careful study he concluded that they were indeed ancient in origin, belonging to a member of the wild, barbarous races encountered by the Roman armies. Schaaffhausen's paper began to draw fire from the anti-Darwinists when it was translated in England. The German was wrong, they declared—the Neandertal bones were thoroughly modern. The bowlegs and low-slung skull were merely the remains of a mental idiot who had suffered from rickets as a child. The poor wretch had spent his life with his brow furrowed against the pain of the disease—hence the protruberant brow ridges. Another critic named Mayer concluded that the so-called Neandertal Man was in fact a Mongolian Cossack who had deserted from the Russian army chasing Napoleon in 1814. The horseman had apparently taken ill and crawled into the cave to die. As "Darwin's bulldog" Thomas Huxley pointed out, however, Mayer's theory failed to account for how or why the dying Cossack would have climbed sixty feet up a vertical cliff and buried his own bones under two feet of loam.

Even when another Neandertal skull was discovered in a small museum on Gibraltar, the anti-Darwinists proceeded undaunted. One of the most eminent scientists in Europe at the time was the German pathologist Rudolf Virchow, and it was he who offered the final word from the generation of scientists who had been trained before Darwin. Ignoring the Gibraltar skull, Virchow scrutinized the disturbing anomalies of the original Neandertal Man and concluded that the specimen was indeed the remains of a man who had contracted rickets as a child—but he had later been subjected to traumatic blows to the head, finally to be victimized by crippling arthritis as an old man! That such a sorry case should survive into old age, Virchow pronounced, was further proof of its recent origin. Only a civilized, agricultural, *modern* society would care for and protect its weaker members to such an extent.

It was easy for the Darwinists to scoff at such desperate explanations—but they were having their own troubles with Neandertal Man. Early in the debate Huxley published a detailed description of the bones. Although they were certainly more apelike in some respects than anything else known, the fossils failed to prove the direct link he sought between the apes and man. From a collection of modern human skulls Huxley was able to select a series with features leading "by insensible gradations" from an average modern specimen to the Neandertal skull. In other words, it wasn't qualitatively different from present-day *Homo sapiens.* Most important, the cranial capacity was every bit as large as that of modern humans—larger, in fact, than those of many living people. Huxley was forced to conclude that Neandertal was old, but fully human. In so doing, he set a standard for judgment—braininess equals humanness—that would have profound implications for the entire subsequent history of human evolutionary studies.

The generation of anatomists inspired by Darwin and Huxley thus had a firm platform of theory to work upon, as well as the obvious anatomical similarities between humans and apes. But for thirty years they had very little fossil material to support their claim of a biological link. Paleontological digs sprouted up in England, Germany, and France. These efforts provided much-needed refinement in methods for dating fossils, but in spite of fervent nationalistic hopes, they failed to reveal the remains of the first Englishman, German, or Frenchman. Then in the early 1890s, a young eccentric Dutchman named Eugène Dubois made a series of discoveries on the distant island of Java. One of Dubois's finds was a skullcap of what appeared to be a manlike chimpanzee; another was a fossilized femur, or thighbone, of an individual that had clearly walked upright. Dubois decided that both specimens belonged to the same individual: an erect-postured, apelike man he dubbed *Pithecanthropus erectus.* In triumph he proclaimed to colleagues in Europe that the missing link had been found.

After the long drought of fossils, it was inevitable that the arrival of Dubois's collection of bones from Asia should trigger enormous excitement. Wherever he lectured he was honored and applauded for his accomplishment. His interpretations, on the other hand, met with more ambivalence. The femur was clearly manlike, but the skullcap was too thick and primitively featured to satisfy the expectations many scientists had for an ancestor of man. Virchow, for instance, dismissed the skull as belonging to that of a gigantic gibbon. Dubois stuck to his guns nonetheless, carting his box of bones to England to be assessed by Arthur Keith, who was beginning to emerge as one of the prominent anatomists of the time.

Keith agreed with Dubois that the specimens could all be the remains of a single individual, but surprised him by concluding that the skullcap should be classified not as an ape, nor as a transitional form between apes and humans, but as yet another primitive human. Echoing Huxley's belief that big brains meant human beings, Keith maintained that the key question was whether or not the specimen had crossed what he called "the cerebral Rubicon." With a cranial capacity of 860 cubic centimeters, *Pithecanthropus* was certainly no mental giant, but its brain was still twice that of some living human beings—microcephalics—who walked upright and could perform simple tasks. Deeply disappointed by Keith, deprived of the confirmation of his "missing link" by nearly all the authorities of Europe, Dubois grew increasingly bitter and vindictive. Legend has it that he finally buried his precious fossils under the floorboards of his dining room. Certainly he restricted access to them.

Before *Pithecanthropus* was whisked out of sight, the view of it enjoyed by Keith and others was enough to convince them that the bones from Java fitted neatly into a ladderlike progression in the human line. First came *Pithecanthropus,* very old and comparatively small-brained, followed by Neandertal Man with his bigger brain but brutal aspects. Finally came true man. Appealing as it was, that simple scenario was about to fall apart, with Keith an instrumental part of its demise. In 1911 he turned his attention to a skeleton found some years before in a place east of London called Galley Hill. This specimen had always posed a problem: It was discovered in deposits which, by the look of the crude tools and extinct fauna found in them, could be dated with some certainty back to the early Pleistocene age. This meant they were much older than Neandertal, perhaps hundreds of thousands of years old. Yet their anatomy suggested a remarkably modern individual. Most authorities before Keith had concluded that the skeleton was the remains of a modern human who had been buried in ancient earth. But Keith decided that "Galley Hill Man" was in fact an ancient progenitor of modern *Homo.*

If the ancient but big-brained man from Galley Hill was our ancestor, where did that leave the more primitive but *younger Pithecanthropus* and Neandertal specimens? Evolution is usually a one-way street; natural selection would not start with something advanced like a big brain, render it more primitive, and then change it back to an advanced form again. Keith's explanation: The human family tree resembled a bush rather than a ladder.

"In the distant past there was not one kind but a number of very different kinds of men in existence," he told a meeting of the British Association in 1912, "all of which have become extinct except that branch which had given origin to modern man." In other words, his (thoroughly British) Galley Hill Man was the true ancestor; the Neandertal and Java specimens were merely the remains of "degenerate cousins."

Belief in man's antiquity was about to gain considerable new vigor—only this time the evidence was to be literally handcrafted. In November of 1912 the *Manchester Guardian* heralded the discovery of an ancient, large-brained protohuman with the jaw of an ape. From what could be understood from the geological context of the fossil, it was hundreds of thousands of years old. Even more exciting, the specimen had arisen not from some Teuton cave or in far-off Java, but right at home in England, in a gravel pit on a farm near Piltdown Common, Sussex.

Three weeks later the discoverers of the fragments—Charles Dawson, a solicitor and amateur fossil hunter, and Arthur Smith Woodward, a leading British paleontologist—presented their evidence to a packed, expectant meeting of the Geological Society. Arthur Keith was there, and so was Grafton Elliot Smith, another of Britain's anatomical bright lights. Both heralded Piltdown Man as perhaps the most important discovery of fossil human remains ever made. But Keith was skeptical about the "apeness" of the skull. Smith Woodward had pieced together the cranium so that the reconstruction had a brain capacity of only 1,070 cubic centimeters—beneath the range of most modern humans. When Keith performed his own reconstruction from casts of the fragments, *his* Piltdown brain turned out to be almost 50 percent larger. Smith Woodward had botched the reconstruction, said Keith—Piltdown clearly represented a much more modern-looking form than its finders had suggested. Nonsense, countered his opponent, who in any case had always thought Keith's insistence on man's antiquity to be "an amusing evolutionary heresy." In the debate Elliot Smith sided firmly with Smith Woodward, elaborating with crisp authority on the unmistakably apelike characteristics of the skull.

That Elliot Smith, probably the greatest neuroanatomist alive, should subject the Piltdown skull to thorough scrutiny and pronounce it apelike is truly astonishing. The skull was merely that of a modern human being. The jaw, on the other hand, was not merely ape-*like:* It was *ape* (an orangutan, to be precise), carefully stained to look old, its teeth filed down to resemble the flat pattern of wear seen in human teeth. The bones had been deliberately placed together in the gravel pit by human hands, so that they could later be "discovered" by a credulous world. Not just Smith Woodward, Elliot Smith, and Keith were fooled by the deception, but the entire British scientific establishment as well.

It would be over forty years before Piltdown was revealed to be a complete fraud. The identity and motivation of the deceiver still remain a complete mystery—Dawson, Smith Woodward, and Elliot Smith are among those who have been suspected of the crime. Most recently, a case has even been put forth to implicate Sir Arthur Conan Doyle, Sherlock Holmes's creator. Living only a few miles away, this master of mystery was fascinated by the Piltdown dig, possessed the chemical and medical talent needed to doctor the bones, and had a grudge of his own against the British scientific establishment. Who knows?

Most accounts of Piltdown leave that question hanging, and quickly move on to what is assumed to be the realy mystery: *How could so many trained scientists be so utterly and shamefully fooled?* But it's too easy to look down on the British evolutionists' gullibility. Consider how little they had to work with. No reliable dating methods. A bare handful of European finds. Some promising Asian material, locked up out of sight by their embittered proprietor. And then along come these sensational new bones, straight out of the British backyard. Of course the English experts would ponder and poke and wrangle over their meaning, little knowing that their very attention to it confirmed the specimen's legitimacy. And of course they would find in the fossil a fulfillment of their preconceived notions: Preconceptions were all they had. Outside of England, especially in America, Piltdown Man caused a great deal more head scratching, and within a couple of decades, enough new evidence would accumulate to reduce the fossil's status from that of a driving force in human paleontology to that of an annoying conundrum. But in the second decade of the century, England was the hub of paleoanthropology, and Piltdown was the evidence. Any new fossil on the block would have to reckon with its implications.

* * *

The fossil whose discovery marks the beginning of modern paleoanthropology came from an unlikely source. Through his mentor Elliot Smith, the young Raymond Dart had obtained an appointment as professor of anatomy at the University of Witwatersrand Medical School in Johannesburg. In spite of its impressive title, the job was no plum. In 1922 the university was barely three years old, located in a glorified gold-mining town composed of little more than tin shacks. It was a very long way, both geographically and intellectually, from the world center of anatomy and medicine that London had become.

Dart felt as if he were being sent into exile. When he and his wife, Dora, arrived at the university medical school, they found the walls of the vast dissecting room pockmarked by tennis balls—a sign of its extracurricular use by the students—and the tables littered with dried-up pieces of corpses. The mind-set of the university authorities appeared to be equally dessicated; Dart's predecessor had been forced to resign in disgrace for having committed the crime of divorce, while Dart himself was regarded as an unfortunate choice as a replacement, simply because he was an Australian.

For eighteen months the couple struggled to keep their spirits up and improve upon the abysmal conditions at the school. Then one day a pupil of Dart's, Josephine Salmons, showed her professor the fossilized skull of a baboon that she had seen on the mantle of a friend, a man who directed a limestone-mining operation at Taung in the Bechuanaland Protectorate. Dart immediately arranged to have any other fossils uncovered by the mining operation sent to him in Johannesburg.

According to his own account, the first shipment of fossils arrived at his house on the very day he and Dora were to host a friend's wedding; Dart himself was to be best man. He postponed dressing for the occasion and ran out to take delivery of two large wooden boxes. The first proved to hold nothing of value, but when he ripped the lid off the second box, he was greeted with the thrill of his life. On the very top was an endocranial cast, the rough impression of a brain that sometimes forms when a fossil skull fills up with debris that later becomes fossilized itself. The rounded stone that Dart held in his hands showed the bumps and crevices of a living brain, forever frozen in time. The cast was clearly from the skull of a primate larger than a baboon.

Dart's neuroanatomical training with Elliot Smith enabled him to recongize immediately a startling feature. Toward the rear of primate brains there are two fissures formed by convolutions in the brain. Elliot Smith had noted that in apes these fissures are close together, while in man they are much farther apart, an indication of the evolutionary expansion of the cerebrum. In this endocast, the fissures were three times as far apart as in any chimpanzee or gorilla Dart had ever seen. Deeper down in the box, he found a large stone with a depression matching the endocast: Faintly visible in the stone were the outlines of the skull and face of the creature.

"Darwin's largely discredited theory that man's early progenitors probably lived in Africa came back to me," he later wrote. "Was I to be the instrument by which his 'missing link' was found?"

While Dart was occupied with thoughts of his place in history, his friend the bridegroom was pounding on the door in impatience. Dart closed up the box, hurriedly affixed his collar and tie, and joined the party just in time to see the wedding come off. As soon as the last guest had left, he returned to the crate. Clearly the next step was to remove the hard limestone matrix, or "breccia," that shrouded the fossil's face. That task would take Dart almost a month. Two days before Christmas of 1924, the rock broke free from the left side, exposing the face of a juvenile individual with a full set of deciduous teeth in place and with the first permanent molars beginning to erupt.

Dart spent the first two weeks of 1925 feverishly writing a report on the find to *Nature,* the prestigious British journal of science. In what has become something of a classic piece of scientific literature, he outlined his reasons for supposing his "Taung child" to represent "an extinct race of apes intermediate between living anthropoids and man." In addition to the position of the brain fissures, the endocast clearly indicated a high, rounded brain quite unlike the flat brains of apes. The forehead rose vertically from the eye orbits without any hint of simian brow ridges. The detentition—though only baby teeth, to be sure—revealed surprisingly humanlike characteristics as well. Perhaps most important for his argument was the position of the foramen magnum, the large hole in the bottom of a skull through which the spinal cord connects with the brain. In apes, which walk on four legs, this hole is located toward the back of the skull. In humans—and in this child from Taung—the foramen magnum is positioned closer to the front, so that the spinal column aligns vertically. This seemed to indicate that the creature had walked erect.

Professor Raymond Dart displays the Taung skull he found in 1924. F. HERHOLDT

Dart proposed for his fossil the ungainly name *Australopithecus africanus*, meaning "southern ape of Africa."

It is not too hard to imagine the shock of the editor of *Nature* when he received this astounding document from a practically unknown anatomy professor, writing from a colonial burg on the other edge of the earth. He responded, like a good editor should, by seeking commentary from the experts in the field—our old friends Arthur Keith, Arthur Smith Woodward, and Grafton Elliot Smith, plus Dr. W. L. H. Duckworth. While the news was filtering through the proper scientific channels in London, however, a reporter in Johannesburg had gotten hold of the story. Dart's "missing link" hit the international press on February 4, 1925. Three days before his *Nature* article was even published, Dart had already received two book offers and congratulations telegraphed from all over the world. The comments of the four English experts were published only a week later, but by this time their measured scientific opinions were muffled in the popular uproar. APE MAN HAD COMMON SENSE! ran the headlines. MISSING LINK THAT COULD SPEAK!

I wonder sometimes how much the instant removal of the Taung fossil to a public arena may have added to its increasingly chilly reception in scientific circles. Certainly Dart's overenthusiasm for speculation did not set the Great Minds at ease about his reliability as a scientist. Keith, Elliot Smith, and Duckworth all published cautious preliminary opinions congratulating Dart, but emphasized the ape characteristics of the fossil over those hinting at humanity. Smith Woodward's review was the most critical, though he joined the others in reserving final judgment until more evidence was forthcoming. In fairness, Dart *had* presented his evidence a little hastily, and with little background information to draw upon in Johannesburg. Moreover, his specimen was a juvenile—and hence did not carry the same weight of telltale characteristics as a mature individual would.

In the critical few weeks that followed, the experts only hardened in their rejection of Dart's claims. Arthur Keith viewed a plaster cast of the fossil. By this time the evidence had reached the public arena in a literal sense, the cast being placed on display at the Wembley exhibition hall. Keith—Sir Arthur, by this time—was obliged to view the cast along with curious members of the public, a circumstance he found extremely irritating. His final judgment?

"[Dart's] claim is preposterous," Keith wrote in *Nature*. "The skull is that of a young anthropoid ape . . . and showing so many points of affinity with the two living Africa anthropoids, the gorilla and chimpanzee, that there cannot be a moment's hesitation in placing the fossil form in this living group."

I am always willing to forgive judgments based on preconceptions when there is little else to go on—but in this case Keith was granted a look at a most compelling bit of evidence for human evolution, and he *still* managed to miss the point. Obviously his condemnation of *Australopithecus* was based on more than an annoyance at having to rub shoulders with the hoi polloi at Wembley. The real problem was that the Taung fossil posed a formidable challenge to Keith's own belief in the antiquity of man. For Keith, Galley Hill Man was still the perfect fossil. (Alas, it later proved to be nothing more than a modern *Homo sapiens* buried in exposed ancient deposits.) Piltdown, with the brain of a man but the jaws of an ape, was at least an acceptable cousin. But Dart's animal showed just the opposite pattern: the brain of an ape but the jaws and teeth of a human. Sir Arthur had only two choices: reject his own carefully evolved theory or pronounce Dart's preposterous.

Dart had shown the further bad sense to have found his fossil in quite the wrong place. By this time Darwin's prediction of an African origin for mankind was obsolete. Darwinism in general had in fact suffered from a sort of ideological atrophy. The tenets of natural selection had been indsicriminately applied in the interests of racism, eugenics, and laissez-faire capitalism. John Scopes was on trial for teaching evolution in Tennessee, and nobody was particularly anxious to find a resemblance between themselves and a chimpanzee. As the birthplace of mankind, Africa—tropical home of dark-skinned degenerates and uncouth apes—was *out.* Asia, cradle of ancient civilizations, was *in.* The influential American paleontologist Henry Fairfield Osborn, for instance, envisioned an Asian backdrop to man's origin as nobly tempered as a Constable landscape. Far from being suited to hairy apes, Osborn wrote, the ancient Asian environment was "a country of meandering streams, sparse forests, intervening plains and meadowlands. Here alone are rapidly moving quadrupedal and bipedal types evolved; here alone is there a premium on rapid observation, on alert and skillful avoidance of enemies; here alone could the ancestors of man find materials and early acquire the art of fashioning flint and other tools."

Through the late 1920s Osborn and others launched expensive fossil-hunting expeditions to China. These American "Missing Link Expeditions" yielded a great number of dinosaur bones, but no missing links.* Raymond Dart, meanwhile, had not followed up on his

*One much more modest Asian expedition did produce a "missing link," of sorts. In 1932, a Yale University graduate student named Edward Lewis returned from the Himalayan foothills with a scrap of fossil jawbone he called *Ramapithecus*. Lewis's claim that his discovery was a hominid was quickly shouted down by far more powerful voices in the science. Discouraged and disgusted, Lewis abandoned paleoanthropology altogether. But thirty years later, his late Miocene jawbone was rediscovered in a drawer by Yale paleontologist Elwyn Simons. Together with his colleague David Pilbeam, Simons convinced most investigators in the 1960s that this species was nothing less than the First Hominid. At the height of its popularity, *Ramapithecus* paraded through a series of Time-Life books as an erect-walking, social, big brained toolmaker—in other words, the quintessential early man. Simons and Pilbeam offered further evidence from the fossil record that the hominid line had diverged from the ape lineage at least fifteen million years ago, and possibly much more.

The legitimacy of *Ramapithecus*'s claim to hominid status was subsequently disputed by biochemists Vincent Sarich and Alan Wilson at Berkeley. They claimed that the very close similarities in some proteins from apes and humans indicated that the hominid line could not have diverged from that of the apes more than five million years ago. If that was true, then *Ramapithecus*, by this time believed to be as old as twelve million years, could not possibly be a hominid. New fossil evidence from the field, as well as subsequent studies showing the near identity in DNA structure between apes and humans, have largely borne them out. *Ramapithecus* is now believed to be an ancestor of the orangutan, and though the accuracy of the so-called "molecular clock" is still under dispute, molecular biology has come to play a critical new role in understanding the human past.

own field with more fossil hunting in South Africa, preferring instead to spend his time developing the neurological and anatomy programs at the medical school. In 1930 he finally got around to completing a detailed monograph on *Australopithecus* and sent it off to the Royal Society in London. The next year he followed up with a visit to London himself, fossil in tow, feeling "confident enough to tackle anything," and sure that in his spiritual home, he could convince his detractors of the significance of Taung.

His timing could not have been worse. Elliot Smith and Keith greeted him warmly, but they were not much interested in hearing his latest thoughts on Taung. Both men were bubbling over with excitement about the recent discovery of Peking Man. After nearly a decade of promise, a quarry in Choukoutien near Peking had finally yielded a magnificent trove of fossils believed to be closely related to Eugène Dubois's *Pithecanthropus erectus.* (The Java and Peking finds were later grouped within the human genus and given the name *Homo erectus.*) The expectations of an Asian origin of man had apparently been fulfilled. In a few days Elliot Smith would be presenting the new finds to the Zoological Society, and he invited Dart along to exhibit his *Australopithecus* skull as well.

At the meeting Elliot Smith launched into a masterful exposition of the significance of Peking Man, replete with projection slides and casts for the audience to handle as he dazzled them with evidence of Peking Man's upright stature, fire-making abilities, and propensity toward cannibalism. It was a tough act to follow. After the applause had subsided Dart was introduced. He appeared at the lectern with only his Taung skull cupped in his hands, and with very little to say that the audience had not already read in print. By the time he was halfway through his presentation, the politely indifferent expressions on the faces before him had shriveled his confidence. His talk trailed off toward its well-known conclusions, and a few days later he left London. Nobody's mind had been changed.

Dart's monograph on *Australopithecus africanus* was never published, and Taung faded out of the public eye. There was no mention of Raymond Dart nor of his find in the most widely read textbook of the time, E. A. Hooton's *Up from the Ape.* Nor did the young Louis Leakey devote a word to Taung in his popular book *Adam's Ancestors,* published in 1934.

Ironically, Leakey himself would be instrumental in bringing the focus on human origins back to Africa. But he too would have his early struggles.

• • •

Louis Leakey's name will forever be linked with Olduvai Gorge, but he was not the first scientist to discover its riches. According to a long-standing legend, on the high plains of the German colony of Tanganyika in 1911, a lepidoperist named Kattwinkel was chasing after a rare butterfly with such rapt attention that he accidentally discovered Olduvai Gorge instead, nearly plunging to his death when the butterfly led him to the brink of the three-hundred-foot drop. Whether or not that story is true, Kattwinkel did bring some mammal fossils back to Germany with him—enough to inspire a legitimate paleontological expedition in 1913 led by Dr. Hans Reck. The enterprise was phenomenally successful. Reck returned with some seventeen hundred fossils, a great deal of information on the geology of the Gorge, and another human fossil to add to the ranks of purported ancestors.

If Reck was right in his assessment, "Oldoway Man" was indeed a prize: a skull and skeleton of an amazingly modern-looking individual, found in deposits he believed to be at least a million years old. But many experts were skeptical, contending that Oldoway Man—like Galley Hill Man—was nothing more than a recent *Homo sapiens* buried by his companions in ancient deposits. To back up his claim, Reck planned another expedition to Olduvai the following year. Then war was declared between Germany and Great Britain, the conflict quickly spreading to their respective colonies in East Africa. The expedition never reached Olduvai, and its exposures lay untouched for over a decade.

Hans Reck would have another shot at defending his Oldoway Man in 1931, thanks to a tirelessly ambitious young man named Louis Leakey. The Kenyan-born son of a missionary, Leakey had formed an early interest in African prehistory and had set his sights on proving that Darwin was right in his prediction of an African origin for man. Though only twenty-eight years old, Leakey had already led two archeological expeditions in Kenya, uncovering some human fossils which he felt bore some powerful resemblances to Oldoway Man. Leakey believed his finds, however, had come from deposits much younger than the ancient sediments Reck claimed to have found at Olduvai. He was all but certain that Reck had been wrong in his assessment of the skeleton's age. And why hadn't Reck found any stone tools on his expedition? Leakey had collected a plethora of stone artifacts from Kenyan sites that he was sure were as old as Olduvai—but Reck insisted there were no tools to be seen at the Gorge. An expedition would clear up these questions, and Leakey invited him along. He even bet the German ten pounds that he could find a stone tool in the Gorge within twenty-four hours of their arrival.

Such spirited self-confidence was typical of Louis Leakey throughout his career. Born in 1903 in a mud house, Leakey grew up among the Kikuyu people that his father sought to convert to Christianity, learning their language before he could speak English and reveling in a boyhood notably free of restrictions. When he was sent away to prep school he found the conventions of polite society almost too much too bear. It was the same at Cambridge University, where Leakey continued to alienate his classmates and scandalize the authorities with his aggressive, boastful manner. He nevertheless graduated with highest honors and greatly impressed his teachers—one of whom was Arthur Keith.

Keith is often credited for planting in Leakey his unwavering conviction in the antiquity of man—a belief that Leakey himself would carry to inspiring heights. But the seed may already have been sown. Deeply religious by nature and training, Louis was still considering following his father's path into the ministry as late as 1925. Leakey was no creationist. But Christianity, unlike many other religions, teaches that man alone is endowed with a soul, and that his dominance over the rest of creation is part of the divine order. It is hard to imagine how Leakey's thinking could *not* have been influenced by a belief in this special relationship between man and the Creator. Such a faith does not easily accommodate the notion of a recent derivation of humanity from a chimpanzee look-alike.

Whatever its roots, Leakey's single-minded quest to prove an early origin for man led him to reject all the candidates that other people had proposed as ancestors. In this he never wavered throughout his long career. Piltdown Man, Neandertal, Java Man, Peking Man, the Taung child—*all* were "aberrant offshoots" to Leakey, the remnants of failed species that probably died out through their inability to compete with "true" *Homo.* The strength of that conviction, fueled by his inexhaustible energy, eventually yielded discoveries that would make Leakey an international celebrity, dragging paleoanthropology into the limelight on the sleeve of his familiar coveralls. But there were decades of frustration first.

The 1931 expedition started off with a bang. The party arrived at the Gorge at 10 A.M. on September 26, and by the next morning

Leakey had found his stone tool, a perfectly shaped hand ax from a period in prehistory called the Acheulean. (Thus in one stroke heralding a magnificent half century of Olduvai archeology, and relieving Hans Reck of ten pounds sterling.) Knowing the Gorge as I do, it is hard to fathom how Reck could have missed finding stone implements on his 1913 excursion—they are lying about all over the place. But it was simply another case of preconception determining the evidence. Reck had been trained in Europe to look for *flint* tools. There is no flint in Olduvai—hence no tools registered in the German's eyes. Leakey, on the other hand, had grown up in Africa and knew what to look for: crude implements made out of quartz, chert, and volcanic rock.

The next round went to Reck. Louis had come to Olduvai almost convinced that Oldoway Man was the product of an intrusive burial. Within four days he had changed his mind, and taken up Reck's position that the skeleton was indeed ancient. Ostensibly it was the presence of very old mammal fossils in the same geological layer where the skeleton had been found that convinced him. But he had been aware of that evidence before he ever arrived. More likely his discovery of an early stone culture tipped the scale in Leakey's mind, never far anyway from coming down on the side of ancient man. In *The Times* of London, Leakey reported that the expedition had established "almost beyond question that the skeleton of a human being found by Professor Reck in 1913 is the oldest known authentic skeleton of *Homo sapiens.*"

Not quite. Within a few months further study of the Olduvai geology and examination of the fossil confirmed that Leakey's first hunch had been right: The skeleton was indeed *Homo sapiens,* but clearly a modern human who had been buried only some twenty thousand years ago in older deposits exposed by faulting. After he had stuck his neck out so far, this could have been a career-crippling revelation for Leakey. But somehow the young Kenyan managed to sidestep the gaff and reemerged with another triumph instead. When Reck left Olduvai to return to Europe, Leakey had continued west to explore some deposits near a place called Kanjera close to Lake Victoria. There he found fragments of two skulls, in sediments containing both hand axes and the remains of a long-extinct species of elephant. Next, at nearby Kanam, a piece of human jaw turned up in what seemed to be even older deposits. Impetuous as ever, Louis concluded that he had found something still older than Reck's Oldoway Man, but with none of the primitive features that damned Piltdown and the Asian finds to dead limbs on the human family tree. He named his discoveries *Homo.*

A few months later, a special conference of learned scientists led by Sir Arthur Smith Woodward—knighted now, like Keith—was convened in London solely to satisfy the experts of the validity of Leakey's claims. On the first day Louis paraded his evidence, and after a second day of laborious discussion, the assembly agreed to his interpretation, congratulating him on what appeared to be a history-making find. But one scientist left the conference harboring some doubts. Percy Boswell, a geologist from the Imperial College in England, was a supporter of Piltdown Man and had been among those who had put the matter of Oldoway Man to rest, so perhaps he was predisposed to look coldly on the claims of "oldest men" from Africa. Whatever his reasons, Boswell picked away at Leakey's evidence with such nagging insistence that Leakey finally requested that the Royal Society provide funds for the geologist to travel down to Lake Victoria and evaluate the sites for himself.

It was an ill-fated trip. When Boswell arrived at Kanjera, Leakey found he could be certain only to within ten yards of the place where he had found the skulls. Kanam was worse. Local fishermen had made off with the iron pegs he had used to mark the fossil's location. A photograph of the site that he had passed around at the London conference proved to be a shot of a place hundreds of yards away. Boswell left the next day, "in a bad humour." Upon his return to England he published in *Nature* a devastating attack on Leakey's claims, concluding that the geological age of the bone fragments from both sites was entirely uncertain.

Nobody was particularly happy about having been misled at the London conference, however unintentionally. Leakey's standing in Cambridge took a further plunge in 1936, when he committed the arch social sin of divorce. He and his wife, Frida, had been spending most of their time apart for two years, and Louis had already begun his relationship with Mary Nicol. They were married soon after the divorce was final. With these two black marks on his record, Leakey would be regarded by much of Cambridge society as persona non grata for the rest of his life.

In the matter of the divorce Louis was right to flout convention: In Mary he had found an entirely more suitable companion than Frida, who much preferred the domesticities of Cambridge to life in the African bush. Leakey's conclusions about the fossils, on the other

Louis Leakey at Olduvai Gorge holding fragments of fossil elephant molars
MELVILLE BELL GROSVENOR/NATIONAL GEOGRAPHIC SOCIETY

hand, were wrong. The Kanjera and Kanam fossils were merely *Homo sapiens.* Leakey himself never accepted this. His entire subsequent career was in a sense a struggle to shore up respect for his Kanam Man with another sensational discovery. Only the next time he would be sure he had covered his bases.

With the Kanam fiasco, Louis Leakey earned in England the same reputation that had shadowed Raymond Dart when he left for Africa ten years before: that of a headstrong junior scientist too caught up in his own ideas to see straight, and too eager to rush to unfounded conclusions. After his own disastrous trip to London in 1931, Dart had returned to South Africa more inclined than ever to devote himself to the medical school and forget about human fossils. Were it not for the tenacious interest of the famous paleontologist Robert Broom, *Australopithecus,* the fruit of Dart's earlier work, would have died on the vine.

Like Dart and Leakey, Broom was brilliant, unorthodox, and unpredictable. In sheer eccentricity he outdid them both. On fossil-hunting expeditions he always wore an impeccable dark suit and starched collar and tie. By the 1920s he had earned an international reputation for his work illuminating the evolution of reptiles into mammals. At the same time, certain lapses and idiosyncrasies—such as his alleged habit of selling national fossils to foreign museums for personal gain—had provoked the wrath of South African authorities. While undoubtedly the most famous scientist in the country, he was nevertheless refused access to its national museum.

Two weeks after Dart had announced the discovery of the Taung child, Broom burst into the young professor's office. Ignoring Dart, he strode over to a table where the fossil lay and flung himself down on his knees "in adoration of our ancestor." By this time Broom was almost seventy years old and had little expreience in human prehistory. Never mind, he later said—he considered himself the best paleontologist who had ever lived, and was determined to become the best anthropologist as well.

It would be twelve years before Broom could act to support his belief in the Taung child's claim to human ancestry. In 1936 Broom got wind that a general store near the quarry of Sterkfontein was selling fossils as souvenirs. He contacted the man in charge of the mining operation, and nine days after visiting the site for the first time, found a skull embedded in rock-hard breccia. Unlike Dart's find, this skull was that of a mature individual. Thus many of the objections raised to the Taung child simply did not apply. Broom recognized the similarities between his fossil and Dart's but chose to call this one *Plesianthropus transvaalensis*—"near man of the Transvaal." Like Leakey, he was a wanton creator of new names. (We call such liberal taxonomists "splitters"; those who take the opposite tack and group as many fossils as possible under one taxon are called "lumpers.") The specimen was later put into the same taxon as Taung, *Australopithecus africanus*.

Broom's next find clearly deserved a taxonomic niche of its own. With the help of a local schoolboy, he located another hominid skull in the nearby quarry of Kromdraai. There was no doubt that it was an australopithecine, but Broom was amazed by the great thickness of the jaw and the massiveness of its teeth. The specimen was overall larger and more robust than either his first find or Dart's Taung skull. Broom gave it the name *Paranthropus robustus*, meaning "robust relation of man." Most scientists now consider it to be a different species of the Taung genus: *Australopithecus robustus*.

With its bulky, overblown features, the new fossil did not appear to Broom to be a direct ancestor to man. But he had no doubt that such an ancestor would be found among the *family* of australopithecines that he had done so much to define—small-brained, erect-walking hominids with humanish teeth. The establishment perked up its collective ear, but remained unconvinced. There was still no indisputable evidence that the australopithecines had walked upright, and the line between them and the apes remained indistinct.

Another problem surrounded the question of age. Before 1960, dating hominid remains anywhere in the world was a chancy business, involving a painstaking analysis of the geology and the fauna found associated with the specimen. In places like Olduvai, the millennia had conveniently constructed "layer cakes" of volcanic ash and sediments, which made it simple enough to establish at least *relative* ages. If you found a fossil in an undisturbed layer of gray ash, for example, you could be sure that it was younger than the fossils in the red layer below, and older than those in the brown layer above. But there were no such guidelines to make sense of the South African caves. They were originally formed when surface water trickling down through the porous limestone dissolved minerals underground, leaving hollow spaces in the earth. As the subterranean caves enlarged, the rainwater might eventually open a channel to the surface, allowing bones and debris to fall in or be dragged in by animals. But when? For how long? It is impossible to say exactly, even today. Dart believed his Taung cave deposits were one million years old, but that was only a hunch. Current estimates put it closer to two million.

By the mid-forties—and with considerable help from Broom and the australopithecine remains he continued to uncover—opposition to Dart's original belief in the hominid status of his Taung skull was crumbling. Even Sir Arthur Keith had to admit there were manlike features in the fossils emerging from South Africa. In 1947, the great Oxford anatomist Sir Wilfred Le Gros Clark swooped down on Broom in Pretoria and submitted all the fossils to his own careful analysis. Le Gros Clark was thoroughly won over. From South Africa he traveled to Nairobi, where he was to be a featured participant at the First Pan-African Congress on Prehistory. The Oxford anatomist used the occasion to put his considerable weight behind the australopithecine claim to human ancestry. With that, the issue was all but settled.

Raymond Dart was at the Nairobi meeting to see his twenty-two-year-old claim vindicated at last. So was Robert Broom, then over eighty years old, but with further discoveries still ahead of him. Sir Arthur Keith, Taung's nemesis, was back in England. When he saw the report in *Nature* he retired to his study and drafted a letter to the journal. "I was one of those who took the point of view that when the adult form [of the Taung child] was discovered it would prove to be near akin to the living African anthropoids—the gorilla and chimpanzee," he wrote. "I am now convinced that Professor Dart was right and that I was wrong."

• • •

The organizer behind that landmark congress in Nairobi had been none other than the indefatigable Louis Leakey. The Kanam debacle long since behind him, Leakey had been working at a furious pace through the thirties and forties, opening up archeological sites all over Kenya and Tanzania. Mary Leakey too threw herself into the work, becoming expert on the stone-tool culture of the early humans of East Africa. Louis preferred looking for bones, but it was Mary who in 1948 spotted their most important fossil up to that point, a skull of an early ancestral ape called *Proconsul.* It was Mary who would make the next big find too, something far more intriguing than *Proconsul.*

The Leakeys had resumed intensive research at Olduvai early in the 1950s. They had uncovered tremendous numbers of mammal bones, including baboons the size of gorillas, and pigs as large as modern hippos. A giant buffalo they uncovered sported horns spanning seven feet. All over the Gorge they saw evidence of early human habitation too—stone tools, assemblages of broken mammal bones that seemed to be the remnants of hominid meals, even an ancient bog filled with the bones of giant herbivores believed to be driven to their deaths by ancient hunters. But the piece that would make the whole picture make sense—not man's leavings, but man himself—continued to elude them.

When the Leakeys finally found what they had been searching for over so many years, it was not at all what they had expected. One July morning in 1959, while Louis lay in bed fighting a bout of flu, Mary took her dogs and walked out to a gully, *korongo* in Swahili, that they suspected held evidence of a habitation by ancient humans. Called FLK, the site had been named years before after Louis's first wife: "Frida Leakey Korongo." Mary noticed a skull eroding out of the slope. Brushing away the dirt, she saw what appeared to be some hominid molars poking out of a jaw. She rushed back to camp to tell Louis, who forgot about his fever and followed her back to the site. What he hoped to find, of course, was a vindication of his Kanam Man—a primitive *Homo* who would account for the crudest of the stone tools of Olduvai. But a look at the teeth left him cold.

"Oh, dear," he reportedly said to Mary, "I think it's an australopithecine."

Leakey's initial disappointment must have been sharp. Through the years he had never accepted Dart and Broom's claim that their small-brained primitives were human ancestors. The fact that his opposition was a voice crying in an increasingly lonely wilderness fazed him not one bit. He was convinced the australopithecines were distant cousins, not nearly human enough to have made the tools found at Olduvai. And yet here one of them appeared to be emerging from the earth only a few feet from a scattering of the very same tools. To make things even more bewildering, when Louis and Mary dug out around the specimen they were amazed to find that the teeth Mary had believed to be molars were in fact premolars. The molars themselves proved to be immense, their chewing surfaces as large around as nickels. Indeed, everything about this skull *except* its brain and front teeth was oversized. Its huge face dwarfed even those found on Broom's "robust" skulls, and a massive bony ridge like that of a gorilla rose from the middle of the cranium to provide extra space for the attachment of the huge chewing muscles powering the animal's jaws.

With this astonishing fossil in hand, Leakey found himself on the horns of a dilemma as wide and unwieldy as those of his giant buffalo. If he called the skull *Homo,* he would have to ignore its obviously robust—indeed *hyper*-robust characteristics. His critics would never let him get away with it. But if he called it *Australopithecus,* then he would have to concede that the South African "near men" were toolmakers, and thus on the line to *Homo sapiens* after all. This would tear apart his whole notion of human evolution. There was a way out that he had used before, but he chose not to take it. Much earlier, Leakey had dismissed the specimens of Peking Man found at Choukoutien as not Man but "relics of his meat feast." Perhaps this creature too was simply an australopithecine who had been eaten by whatever true man was making the tools. But he rejected that idea. The skull had been fossilized while still virtually intact, and there were no tool marks anywhere on it. Besides, this skull was a *Leakey* find. He would never have admitted it, but that fact alone made it a little more human.

Leakey's field notes, which I came across when I was in Nairobi in 1980, bear witness to his consternation. In his journal, he made a list of the robustlike characteristics of the fossil, and another list of those traits that aligned it with *Homo.* The robust list was far longer than the *Homo* one. Nevertheless, in the journal he identified the fossil as *Titanohomo mirabilis,* or "miraculous giant man."

"He is a fabulous creation," Leakey wrote, just below a rough sketch of a top view of the FLK skull. "*Titanhomo mirabilis* would be a good name. People will say he is *NOT* human but he is."

A page from Louis Leakey's field diary after the discovery of "Zinjanthropus" in 1959. "He is a fabulous creation," Leakey wrote. "'Titanohomo mirabilis' would be a good name." DONALD JOHANSON

When it came time to publish, however, Leakey dodged the dilemma rather than resolve it. Rejecting both *Homo* and *Australopithecus,* he chose instead an entirely new name for the fossil, *Zinjanthropus boisei.* The genus name meant "East African Man," while the species designation honored Charles Boise, a generous donor. For a nickname Louis and Mary called it "Dear Boy," and another scientist gave it the sobriquet "Nutcracker Man," because of its powerful jaws. But the nickname that stuck was simply "Zinj."

Whatever he called the thing, Leakey was explicit about what it was. In his report to *Nature* he quickly acknowledged the many characteristics it shared with the australopithecines, even placing it within the same subfamily. But then he began to pull *Homo*-like traits from the skull too. His conclusion was that Zinj and the toolmaker were one and the same. After a quarter of a century, Kanam Man had at last found an ally.

Zinjanthropus made the Leakeys' name virtually a household word around the world. It also brought them the support of the National Geographic Society, and an end to the constant financial worries hounding their research. But the notion of a relatively small-brained hominid with millstone molars and the bony crest like that of a gorilla atop its skull did not sit well with many scientists. Leakey had another problem. Estimates of the fossil's age put it at about 600,000 years. According to his own notions on the speed of evolution, this did not allow nearly enough time for this primitive East African Man to develop the anatomical refinements of *Homo sapiens.* Louis explained away this dilemma with typical aplomb: Everybody knows that domesticated animals evolve faster than animals in natural habitats. By developing a tool culture, wasn't man in a sense hurrying up his own evolution by *domesticating himself*?

Possibly—but you can never really put such a subjective idea to the test. Leakey had a knack for weaving arguments together that could make far more preposterous notions ring with truth. In this case, however, his rhetorical skill would not be needed. Almost immediately after the discovery of *Zinjanthropus*, a new chemical method of dating volcanic deposits such as those at Olduvai came up with an objective, verifiable age of 1.75 million years for the FLK site.

Known as potassium-argon dating, the technique was to revolutionize the science of paleoanthropology. It took advantage of the fact that radioactive isotopes in volcanic rock—in this case potassium-40—"decay" into stable compounds—argon gas, in this case—at a steady rate over millions of years. By taking a sample of volcanic crystals and measuring the proportions of radioactive potassium to trapped argon gas, geologists Garniss Curtis and Jack Evernden at Berkeley were able to come up with absolute ages for the volcanic samples. The technique required incredibly sensitive instrumentation, and especially in its early forms, it was very sensitive to error by contamination from rock crystals that might have become mixed into a tuff from other geological strata. It could never be applied, furthermore, to date fossils directly—only the deposits that surrounded them. In spite of these limitations, however, potassium-argon dating provided a longed-for objectivity to a science overloaded with subjective interpretation. In time, the technique would prove to be a more priceless discovery than Zinj himself.

Leakey, of course, embraced the new age of FLK wholeheartedly. Why not? In one stroke his East African Man had been given more than a million extra years in which to evolve. With his customary flourish, Leakey told the readers of *National Geographic* that he had finally uncovered his first true man, the maker of the tools at FLK, "a state of evolution nearer to man as we know him today than to the near-men [i.e., *Australopithecus*] of South Africa." Most of his colleagues, however, were inclined to believe that his first impressions of the skull's identity had been right: It was another, hyper-robust type of australopithecine. Zinj eventually wound up in the same genus, with the formal name *Australopithecus boisei*.

Within a couple of years, news began to trickle in that Leakey had turned up something else at Olduvai, practically right under Zinj's feet. In November 1960, just two months after publication of the

National Geographic article, Louis and Mary's oldest son, Jonathan, was excavating the find of a saber-toothed cat near the FLK site when he uncovered a piece of lower jaw, or mandible, and some cranial fragments belonging to a juvenile hominid. Found in deposits the same age as those surrounding Zinj, the skull bones and teeth of "Johnny's Child" showed none of the merely near-mannish qualities of the previous find. Most important, it bore every sign of a big brain.

Johnny's Child or OH (Olduvai Hominid) 7 as it is officially called, was in fact far more like the true *Homo* Louis had been searching for all his life. But this time Louis swallowed his usual impatience and withheld announcement until the evidence could be studied in depth. With astounding good fortune, the Leakeys' survey team at Olduvai meanwhile found two more skulls bearing similar traits. The first was OH13, an upper and lower jawbone and some cranial fragments. This specimen was given the name Cinderella, or Cindy for short. Soon afterward OH16 was discovered—a trampled cranium and some teeth—near FLK. This one they dubbed George. These skulls echoed the humanlike traits of Johnny's Child. The cranial capacities were large. The cheek teeth—molars and premolars—were narrow, and small relative to the canines. Some hand, foot, and other postcranial bones had also been found at the OH7 site. The hand bones showed a finely developed precision grip, such as a tool user might possess. And the bones of the foot showed evidence of a full bipedal gait.

These traits were more than enough to convince Louis that the new hominids were unqualifiedly *Homo*. But before he could announce a new species, Leakey had another obstacle to overcome. Oddly enough, the genus *Homo* had never been clearly defined in the first place. Most investigators judged the humanness of a fossil skull on the basis of whether it had crossed Arthus Keith's "cerebral Rubicon." Just how big a brain had to be to pass over this famous threshold varied from between 700 and 800 cubic centimeters, depending on who was doing the judging. According to Leakey's colleague Phillip Tobias, the cranial capacity of the new Olduvai specimens averaged around 650 ccs. If they wanted to call them *Homo*, they would have no choice but to declare that the cerebral Rubicon was lower than anyone thought. They nudged it down to 600 ccs, which put their new fossils on the human side. Joined by John Napier, Leakey and Tobias announced their new species in a landmark paper in *Nature* in 1964. Here they said, was the *real* toolmaker of

Olduvai Gorge, *Homo habilis,* or "Handy Man," a name suggested by Raymond Dart. *Zinjanthropus,* never a very convincing human ancestor in the first place, fell in Leakey's estimation to become "an intruder (or a victim) on a *Homo habilis* living site."

In spite of Leakey's careful preparation this time, the announcement of his new *Homo* received a cold reception. On this occasion even the great Wilfred Le Gros Clark, long a Leakey friend and champion, deserted him. In essence, the paleoanthropological community had heard Louis cry "Man!" one time too often. Some scientists were uncomfortable with the fiddling necessary to squeeze *habilis* across Keith's cerebral Rubicon. To make matters worse, the specimens that Tobias had measured were incomplete and badly broken, and some of his colleagues questioned the accuracy of his measurements. On the basis of the teeth, Leakey and his two co-authors were on firmer ground. Using Le Gros Clark's own distinctions between *Homo* and *Australopithecus* teeth, the *habilis* dentitions fell into the former genus.

But it was not enough to win over many converts. Taxonomic "splitting" was out of fashion in the 1960s; most investigators were looking for ways to *reduce* the number of players on the hominid scorecard, not add new ones. Le Gros Clark summed it up for the opposition: All the new specimens from Olduvai Gorge could just as easily be accommodated into Dart's species *Australopithecus africanus.*

"One is led to hope that 'Homo habilis' will disappear as rapidly as he came," he wrote. "It certainly does not appear to merit prolonged controversy."

The most damning element in that response is the quotation marks around the name of the new taxon—little whisk brooms to brush aside its claims to the legitimacy conferred by italics. But in fact *Homo habilis* sparked one of the longest and most prickly controversies in the history of the field—one that would outlive Louis himself and is still very much unresolved. But I'll save that one for later. "Handy Man" was Leakey's final great find; he spent the last decade of his life involved mostly in lecturing and raising funds. He and Mary grew further apart. While Louis was regaling audiences around the world, Mary dug in at Olduvai Gorge. Eventually she would make it her permanent home, coaxing from its karongos an astonishing wealth of information on the stone culture of early man. Meanwhile, the paleoanthropological spotlight shifted its focus north, to fall upon Richard Leakey's finds near Lake Rudolf in Kenya and upon our own discoveries in the Afar of Ethiopia.

PART II:

The Road to Olduvai

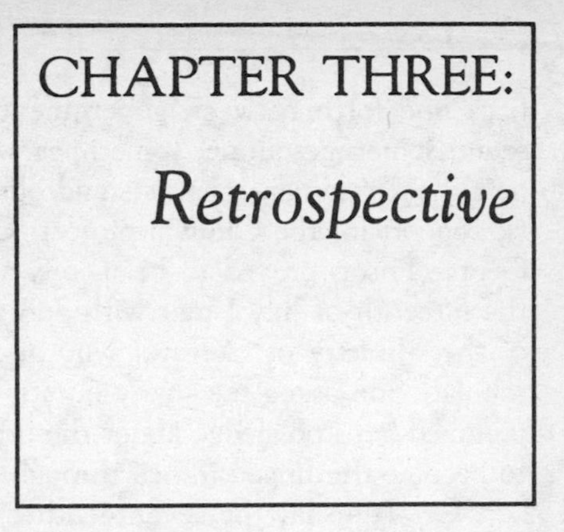

CHAPTER THREE: *Retrospective*

Science must begin with myths, and with the criticism of myths.
—Sir Karl Popper

DAR ES SALAAM IS NOT a beautiful city, but I still enjoy walking its streets. The town has a well-used, frayed look to it, rather like an old pair of shoes that you keep around out of habit and obstinate affection. Built by a sultan of Zanzibar in the 1850s, Dar's fortunes have been scoured by tides of occupation by Arabs, Portuguese, Germans, and finally the British. After independence was declared in 1961, the city blossomed as the capital of the new country of Tanzania, which combined the former colonies of Tanganyika and Zanzibar off its coast. Colonial buildings were converted to civic use, highrises and factories sprouted up, new warehouses and docks cluttered the shore to service trade in the port. But in the 1970s, the development began to lose momentum. In 1985, when I had passed through on our brief excursion to Olduvai, there were signs all around of improving conditions. By the following summer, the capital seemed even more invigorated. Whatever its economic fortunes, Dar never seems dreary or depressed—it is too full of bustling life. I've grown fond of its people, its crowded sidewalks and quirky shops, and the way the city seems to shake the dust off its streets every morning and start over.

For three days following my sleepless night in Dar, I shuttled back and forth between government offices and the National Museum, sometimes alone, sometimes with Fidelis Masao. For the most part the Tanzanian scientists and officials reaffirmed their enthusiastic support for the Olduvai project. Clearance for Lewis Binford and George Frison proved to be merely a matter of writing a letter. On the fifteenth of July I met with Zahra Nuru, the Principal Secretary of the Ministry of Culture, who also turned out to be a linguistic scholar. She asked me many questions about anthropology, while I plumbed her knowledge about the origins of Kiswahili and its spread to become the lingua franca through much of East Africa.

On Thursday the seventeenth I left Dar for the regional capital of Arusha, the research permit for Olduvai tucked safely in my pocket, along with clearances for all scientists and visitors. I had assurances from officials in the Department of Antiquities that the permit to excavate in Laetoli would soon follow. But I was still nagged by doubts. Earlier, I had learned that Richard Leakey had passed on to an official in the department a rumor that we had found two hominid skulls at Olduvai on the scouting expedition the summer before, and had taken them out of the country.

I had heard that Leakey was very angry about our "claim jumping" into his mother's territory, but I was still puzzled and dismayed by this report. It seemed impossible that Richard, with all his experience, would lend credence to such a notion. In all of its glorious history, Olduvai had produced only one complete skull—Zinj—along with a handful of crania like Cindy and Poor George. How could we have located two skulls in a period of ten days? And even if we had made such astounding finds, why in heaven would we try to smuggle them out of Tanzania? Anthropological contraband is useless. You cannot publish anything about it without exposing your theft, but if you don't publish, the material cannot be considered evidence—in effect, it does not exist. Whoever was responsible for the rumor, it indicated that there was still some suspicion of our motives. I would not feel wholly comfortable until all permits were secure.

I checked into the New Arusha Hotel in the early afternoon and called Sandy Evans, the local representative for Abercrombie and Kent, a prestigious outfitter for tourist safaris in East Africa. Through my friendship with Geoffrey and Jorie Kent, I had been able to contract with A&K to provide transportation and a supply line to the Gorge. Sandy told me that Gerry Eck, Lew Binford, and several other members of the expedition had decided not to wait for us and had departed for Olduvai that morning. I couldn't blame them.

Reaching the field becomes an overwhelming urge even for seasoned investigators, growing harder to resist the closer one gets to the destination. I only hoped that Gerry had loaded up on supplies before leaving. The A&K vehicles that would be ferrying us in and out could easily handle the rough road up the Ngorongoro Crater, but Sandy Evans could give me no assurance that supplies of vegetables, flour, and other basics would be routinely available.

Through the afternoon the remaining members of the team gathered at the hotel. Tim White and Berhane Asfaw flew in from Addis Ababa. Gen Suwa, a Japanese graduate student working with Tim, had come in by way of Nairobi. Unfortunately, Gen's luggage did not arrive with him, and he was still out at the airport trying to track it down. Tim and I walked over to the uncompleted National Natural History Museum, housed in what had been the German administrative headquarters for the region before Great Britain took control of the colony after World War I. We hoped that Pelaji Kyauka, the museum's Principal Curator, would be able to join the expedition for a few days; his Land-Rover would be especially useful in transporting the group to Laetoli. Kyauka was eager to go but had too many commitments to leave immediately; if all went well he would follow us out on Monday. When we got back to the New Arusha, Gen Suwa had checked in, looking very ragged and still without his luggage. We were all bushed and retired early.

At seven o'clock the next morning a sleek beige passenger van was waiting for us at the door of the hotel, ABERCROMBIE AND KENT painted in swirling chocolate-colored script across its sides. There was still no sign of Gen's bag; all he had with him was a toothbrush, a camera, a couple of books, and the clothes on his back. It looked as if he would have to stay behind in Arusha, then hitch a ride out with Kyauka on Monday.

"If you leave that suitcase at the mercy of the system, you might never see it again," Tim said over breakfast. "Wait here and get it cleared up. You'll be out in the field soon enough."

Tim has a way of putting things that make it difficult to contradict him, especially if he is your graduate-thesis adviser. But Gen looked ready to burst. After six years of study, poring over specimens in museums, reading paper after paper, the thought of three days cooling his heels in Arusha while the rest of us were trudging over the slopes of the most famous fossil site in the world was obviously a crushing prospect.

"There's nothing in that bag I can't get along without," he muttered into his plate. Tim shook his head and laughed.

"Okay, you win," he said. "Now does anybody want to lend Mr. Suwa a pair of socks and some underwear so we can get out of here and go find some damn hominids?"

We all pitched in some spare clothes to the "Gen Fund" and left his suitcase to its fate. Outside we piled into the waiting van—one of those eight-passenger affairs called "combis" that have been specially customized for wildlife sightseeing. With our blue jeans, T-shirts, and cameras we could easily have been taken for a rather motley group of tourists headed out on safari. Each of us sank into a plush swiveling bucket seat. We made a few stops for bread and other perishables, and then we were finally on the road, heading west out of town. The storefronts soon gave way to a messier conglomeration of long, low industrial buildings, junkyards, and dusty lots, interspersed with greener patches of vegetables. Gradually the green took over. Along both sides of the road groups of women in bright-colored tunics walked in toward town, chatting to each other. Beyond, rows of coffee bushes and banana trees marched away from the road. We passed over several streams, their banks crowded with women washing clothes and children splashing in the water.

The combi purred along, bouncing gently on cushioned shocks. It was an absurdly stylish way to travel to a dig, but before long I decided to stop feeling silly and just enjoy the ride. Out the window, the cultivated greenery gave way to a plain of low yellow grass, pockmarked here and there by isolated rounded hills—the eroded remnants of volcanic cinder cones. As quickly as signs of water disappeared, the human population thinned as well. Soon we were passing only an occasional pedestrian or bicyclist. In the distance I could see a line of cattle moving across the plain, Maasai herdboys lingering by their flanks. For the first time I felt wholly in Africa again, a moving point in an open plain studded with termite mounds and flat-topped acacia trees. Above, the blue vault of the sky straddled the horizons, the entire continent held in its grasp.

When I first set foot in Africa it was under much different circumstances. I was twenty-six at the time, just another graduate student in Clark Howell's lab at the University of Chicago, with nothing more substantial to offer than a little training in archeology and a growing familiarity with the arcana of ape teeth. Having received a small travel grant to Europe to gather data for my doctoral dissertation on chimpanzee dentition, I asked Clark if he would let me extend the trip and join his field expedition in East Africa. To my everlasting joy he agreed.

For three years Howell had been looking for hominid fossils in the Omo of Ethiopia, a one-hundred-square-mile, fossil-rich region named after the river that meanders through it before spilling into Lake Turkana, just over the border in Kenya. The hominid specimens Clark had discovered in the Omo before I arrived were themselves not exceptional, comprised mainly of teeth, jaws, and bone fragments whose telltale surface features had been "rolled"—worn smooth by fast-moving water. In fact, Clark would never make the big find that he hoped for in the region. What the Omo did provide was a uniquely legible record of mammalian evolution over a period from four to one million years ago. Ancient pigs, antelopes, elephants, giraffe—all evolving from the forms found in the older horizons to the forms present in younger one.

These mammal fossils, especially the pigs, proved to be the Omo's ace in the hole. By cataloguing the morphological changes in the pigs found through successive strata, paleontologists could trace the appearance and disappearance of pig species as they flouished and died out through the millennia—in effect, watch evolution proceed. Then, by correlating the species' times of origin and extinction with absolute ages provided by the potassium-argon dating technique, the evolving family lines of pigs themselves became a measure of the passing of time—not just in the Omo, but wherever those pigs could be found, the great gift of the Omo was thus a second yardstick, a way to assign ages to deposits that did not contain volcanic ash and were therefore beyond the reach of the potassium-argon technique. In one case, the Omo's "ace in the hole" would trump a claim for the oldest man ever found.

For me the Omo was a glorious, if sometimes painful, awakening. Like every rookie in the field, I harbored a secret conviction that I would be graced by an impressive find of some sort. Out there was a hominid jaw or skullcap lying in the dust, waiting for Johanson alone to come and complete its destiny. On the flight from Nairobi to the Omo expedition landing strip, I rehearsed the scene over and over in my mind—I saw myself rushing on ahead of the other surveyors, drawn by some magical affinity to the spot where a row of shining teeth was just beginning to erode out of the soil. "Over here! I think I've got something!" I'd shout. Clark Howell and the rest would come running over, flushed and out of breath—"*Good God, the kid's found a perfect mandible. . . .*"

What I really discovered in the Omo was my own ignorance. For my dissertation I had already amassed data on every nook and cranny of a chimpanzee's every tooth, and along the way had devised a new system for comparing dental characteristics of all the apes, including the extinct hominids. But I was pretty naïve about doing paleoanthropology in Africa. Minutes after the plane set down on the expedition landing strip, I was unpacked and out in the field, looking for fossils. That part of my daydream, at least, was coming true. Just as I'd imagined, the ground was practically littered with bone. The problem was that I had no idea what many of the bones were—which were antelope, which were pig, much less which were possible hominid. I also found myself stupified by the heat, on the verge of passing out. Far from bounding on ahead of the group, I was struggling to keep up.

At Clark's suggestion, the next day I stayed in camp in the afternoon, helping him sort through the bags of bones that had been collected from the morning's work. Over the next few weeks I gradually learned to recognize identifying features of the specimens. I also learned how to make myself useful at the routine, apparently trivial tasks essential to keeping an expedition alive. I mapped sites, sorted supplies, fixed vehicles, and helped keep the camp well-fed and happy. In short, I was learning what it takes to organize a field expedition, and after two more seasons in the Omo, I'd become fairly good at it.

In the early 1970s, research in the Omo was mounted jointly by French and American teams, their camps separated by about one mile. The two teams had different research approaches, and the relationship between them was sometimes rather distant and cold. I grew curious, however, about these other scientists living parallel lives so close to our own base of operations, and after a while I took to visiting the French camp, eventually making friends among the staff. That little bit of sociality turned out to have enormous consequence. The connections I made in the French camp led me to a young geologist named Maurice Taieb. Taieb had begun a study of a part of the Awash River valley in the Afar Triangle, a region northeast of Addis Ababa named for the Afar tribe indigenous to the area. He claimed that the deposits looked to be three million years old, and there were fossils lying about all over the place.

"Why don't you come out next year and see for youself?" he said. He made it sound so simple.

I took Taieb up on his invitation, but it was far from simple. Clark Howell objected that I still hadn't finished my doctoral dissertation, and that I had just accepted a teaching position at Case Western Reserve University in Cleveland, and had no business gallivanting around looking for new field sites. And where was I to raise the money for such an expedition? I knew Clark was right, but I still couldn't resist the itch to explore unknown territory. I was eager to get out on my own, and the Afar sounded like the right place to go. I was encouraged in my boldness, moreover, by a clear precedent: Several years earlier, when only twenty-three, Richard Leakey had himself left the Omo expedition and taken off on his own. Richard set up camp at Koobi Fora, a promontory on the eastern shore of Lake Rudolf (now Lake Turkana). When I first met him in Nairobi in 1971, he was already well-known for the fossils he had uncovered at Koobi Fora, in particular a beautifully complete *Australopithecus boisei* skull much like the Zinj fossil his parents had found at Olduvai ten years earlier. If the Afar held anything half so valuable, I wasn't going to wait around for someone else to find it. I managed to scrape together a little grant money; combined with my savings, it would be enough to finance a short exploratory mission to the Afar with Taieb.

In the spring of 1972, before the Omo season was to begin, Taieb and I met in Addis, where we were joined by geologist Jon Kalb. When we reached Hadar, I could see immediately that the gamble had paid off. Just as Taieb had promised, the Hadar deposits were riddled with bone, beautifully preserved mammal fossils dating—from the look of the pigs—from about three million years ago. After three months in the Omo, I flew back to the United States and took up my teaching job, working feverishly through the nights to complete my dissertation. Once I was over that hurdle, I applied to the National Science Foundation for funds to support an expedition to Hadar in 1973. The NSF awarded me $43,000. I wasn't a proven investigator yet, but I was no rookie anymore either.

Fourteen years had passed since then—the most tumultuous period of discovery in the history of my science. Now I was back in Africa once more, the sun-drenched grasslands speeding by on both sides as the combi rolled down the tarmac. I had first traveled that road on my way to visit Olduvai Gorge in 1972 as a guest of Mary Leakey's. I knew Mary only slightly, and I was nervous. She had a reputation for being difficult to get along with; one scientist I had talked to claimed she had thrown him out of camp for having "insulted" one of her dogs by helping it get into the back of a truck. But whatever her reputation, she proved to be invariably friendly toward me in those early days. She liked having someone around who made her feel comfortable, and she seemed to respect the risks I was taking to explore unknown Hadar.

"You were right to get out of the Omo," she told me one evening while we watched the twilight arrive over the Serengeti. "Louis and I told Clark Howell that the Omo was no good, but he wouldn't listen. But you'll find something. You're like a Leakey."

In the fall of 1973 the International Afar Research Expedition established its first camp on a level bluff overlooking the Awash River—a location that provided both water and a little relief from the furnacelike heat of the canyons and gullies where the surveying would actually take place. Taieb plunged headfirst into the task of figuring out the complicated Hadar stratigraphy. The rest of us followed suit. Day after day I scrambled around in the stifling heat. But if there were hominids to be found, they were keeping themselves well hidden. After several weeks of exploring, I had exhausted most of a grant that was supposed to have lasted two years, and had nothing to show for it. I wondered how I was going to explain that to the NSF.

Then, near the limit of our time in the field, Hadar suddenly fulfilled its promise. While surveying late one afternoon with a colleague named Tom Gray, I uncovered what looked to be a monkey's proximal tibia—the top end of a shinbone. A few yards away I noticed a distal femur—the lower end of a thighbone—lying in two pieces on the ground. I put the two pieces of femur together, and then carefully fit them against the tibia. That they matched perfectly was not surprising, since they were the same color, lay next to each other, and thus might be expected to form the knee joint of a single individual. What was surprising—astonishing, in fact—was the *way* they fit together. The thigh and shin bones of monkeys join to form a straight line. These two bones met at an angle, the femur slanting outward. There was only one living primate endowed with such a knee joint. Human femurs angle outward in order to give balance for walking on two legs. I could scarcely believe the evidence in my hands. If our preliminary dating of the Hadar deposits was correct, I was holding the knee joint of a hominid over three million years old: the earliest record of a bipedal ancestor yet discovered.

What *sort* of hominid I had found was anybody's guess. It was impossible to declare an isolated knee joint *Homo, Australopithecus,* or anything else; all I could say for sure was that the creature had been bipedal and surprisingly small. Most of the features on a skeleton that can pin down its identity are found from the neck up. Australopithecine faces, for example, project forward more than human faces do, and their jaws are generally more robustly constructed—presumably better suited to grinding up a diet of tough vegetable matter. Their molars are correspondingly larger, while *Homo*'s back teeth are small relative to their front teeth. And then there is the question of brain size. In the mid 1970s, a large brain was still considered such a hallmark of humanity that for many scientists, a specimen exhibiting that one feature alone would be pronounced human—regardless of less portentous traits that might argue otherwise. What I hoped to find when I returned the next year was a skull to go along with my knee joint, or at least enough of one to tell me what I was dealing with.

The year before, one of Richard Leakey's skilled team of fossil finders had come across just such a find at Koobi Fora. After Richard's wife, Meave, and anatomist Alan Walker had painstakingly reconstructed the skull from hundreds of fragments, it proved to be that of a hominid with a cranial capacity of 775 cubic centimeters—certainly larger than any australopithecine known. But what was most remarkable about the skull was its apparent age. On the basis of a potassium-argon date given to a volcanic tuff above the deposit where the skull had been found, Richard believed it to be 2.9 million years old. Such a date had stunning implications. The oldest known australopithecine was nearly half a million years *younger.* If the skull was *Homo*—and Richard emphatically believed that it was—then Taung and the other australopithecines could be dismissed from the human line after all. Richard had found the Oldest Man. Apparently, Louis Leakey's claims for the antiquity of our ancestry had been vindicated by his son.

By this time, Louis had been in bad health for several years, battling arthritis and a heart condition, refusing to relinquish responsibilities and curtail his schedule of public lectures. His relationship with his second son had long been strained, particularly since Richard's early successes. But when in September Richard brought the new skull from Koobi Fora to show his father, Louis was overjoyed. He believed the fossil—known to the world as KNM 1470, its acces-

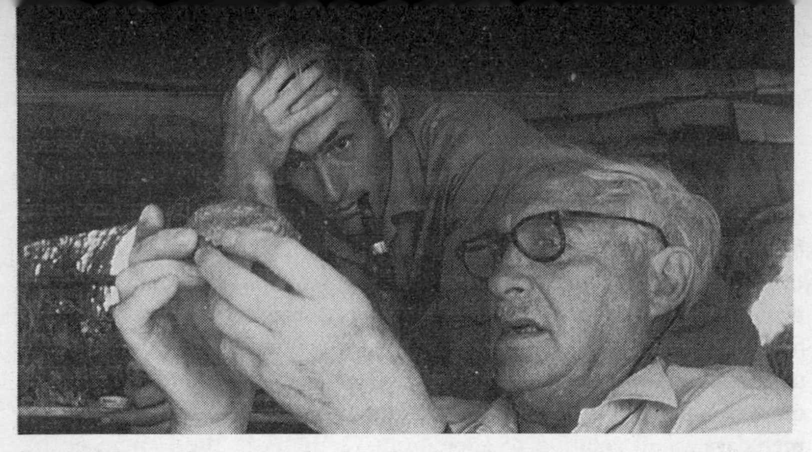

Louis Leakey and his son Richard study a fossil cranium together at Koobi Fora, shortly before the elder Leakey's death. Richard's discovery of the Homo habilis *skull 1470, which he believed to be close to three million years old, appeared to vindicate his father's belief in the antiquity of* Homo. GORDON W. GRAHAM/NATIONAL GEOGRAPHIC SOCIETY

sion number in the Kenya National Museum—to be the final confirmation of *Homo habilis,* the embattled species he had named almost a decade before. With 1470 in hand, Louis's lifelong search for the true early *Homo* had ended. That same evening, Richard drove his father to the airport to catch a plane to London, the first leg of yet another grueling lecture tour of the United States. A few days later, Louis died of a heart attack in a friend's apartment in London.

The 1974 expedition to Hadar would yield its own share of discoveries. First came a series of remarkable finds by Alemayehu Asfaw, an Ethiopian member of the team: three beautifully preserved hominid jaws, two of them found within an hour of each other. One of the local Afar workers turned up a fourth jaw only a few days later. Now we had something more to go on, something with diagnostic possibilities. That the jaws were hominid and not ape was certain—their canines, to take just one example, were much less pronounced than those of gorillas or chimpanzees. But deciding whether the jaws were *Australopithecus* or *Homo* was much harder. Some of their characteristics were clearly primitive. One, for instance, showed a gap between the upper canine teeth and the incisor teeth immediately adjacent—an accommodation for projecting canines in the lower jaw. On the other hand, their molars were smaller than any I'd seen on the robust australopithecines from South Africa, and in general their jaws possessed *Homo*-like proportions. One of the jaws seemed as if it would fit very nicely onto Richard's 1470 skull. But were they really *Homo*? A reliable judgment, if one was there to be made, would require careful study back home in the lab.

In the meantime, it wouldn't hurt to have some other experts take a look at the jaws. I was itching to show them off, hear some opinions. Not everyone works this way. Some of my colleagues like nothing better than to hunker down over an enigmatic bone in solitude, absorbing its every detail before discussing it with other scientists. I am a throw-it-on-the-table type. My best insights are inspired when fossils are passed hand to hand, turned to catch the light, compared, argued over. I decided to write Richard and invite him up to camp.

Richard and his wife, Meave, flew in from Nairobi on November 28, together with Mary and paleontologist John Harris. To my delight, the Leakeys agreed with my suspicions that the jaws might be *Homo.* Richard and Mary were not only my friends, they were also the most famous investigators working in East Africa. Their opinion carried a lot of weight with me. If the jaws were indeed *Homo,* then I had found in the Afar no less than the oldest trace of man's tenure on the earth. That is a seductive idea for any physical anthropologist. For someone as young and ambitious as I was, it was almost irresistible.

After seeing my visitors off on the morning of November 30, I decided to remain in camp and begin planning a paper that would describe the jaws and tentatively assign them to the genus *Homo.* That was the morning that Tom Gray came into my tent and asked if I could show him a section of the site named Locality 162. A few hours later Lucy arrived in my life.

Before leaving camp that morning I'd sensed that we might find something significant. I seldom get that feeling, and when I do it usually pans out. But after surveying for several hours at Locality 162, we had uncovered no more than some horse and antelope teeth, a fragment of a pig skull, and a piece of monkey jaw. None of this added anything new to our overflowing mammal collection. It was close to noon and there was not much point in continuing. Most fossils are discovered in the early morning or late afternoon, when the oblique rays of the sun lend shadow and definition to objects on the ground. As the sun climbs higher in the sky, the gravel and rocks and bone all seem to fade into each other. At midday you could be staring right at a perfect arcade of teeth and not see a thing.

We headed back to the Land-Rover through a little gully on the other side of a rise. As always, I kept my eyes moving along the ground with every step. I knew that the gully had been worked over a couple of times before, so I wasn't surprised when it appeared to be empty of bones. But just as I turned to leave, I saw what appeared to be a fragment of an elbow joint lying at the bottom of the slope above. Tom and I knelt down to examine the thing. It was small, very small, but unquestionably a hominid. Then I spotted a piece of skull next to Tom's hand, and suddenly we seemed to be surrounded by hominid bones—a femur, a piece of pelvis, ribs, some vertebrae. For a while we just groped around from one bone to the next, too stunned to speak. It occurred to me right away that perhaps all these bones might belong to a single individual. But I was afraid to speak that thought out loud, as if by doing so I would break the spell, and we would once again find ourselves standing in an unremarkable, boneless little gully in the middle of nowhere. Tom, on the other hand, could not hold in his excitement. He let out a yell, and then I heard myself yelling too, and we were hugging each other and dancing up and down in the heat.

As we brought the fragments of her skeleton back to camp over the next few weeks, we laid them out as they would have been in life—vertebrae in line, rib fragments branching out in parallel arcs, the top of the thighbone firmly nestled in its pelvic socket. The effect was uncanny. We all shared the feeling that this ancient creature was being re-created, coming to life before our eyes. I knew as well that as Lucy emerged, my own life was irrevocably changing. Whatever she turned out to be, Lucy was sure to be one of the biggest finds of the century. To point to another skeleton in anywhere near the same state of preservation, one would have to advance in time all the way to the Neandertal fossils from the middle Paleolithic period in Europe, 75,000 years ago. This one was easily forty times as old. If I had made a significant find with the knee joint the year before, with Lucy I had made a reputation.

She did nothing, however, to clarify the identity of the existing Hadar collection. I had been leaning toward calling the jaws *Homo,* but the new skeleton brought me up short. The pieces of mandible recovered from the Lucy site told us that her lower jaw was V-shaped, a decidedly more primitive condition than anything associated with *Homo.* Her first premolars, the teeth directly behind the canines, had only one cusp as opposed to the two found in *Homo*—another primitive trait. The cranial fragments meanwhile suggested a very small brain, not much larger than that of a chimpanzee; indeed, at three and a half feet tall, Lucy was about the size of a chimp. Much as I wanted to believe that Lucy was human, I couldn't bring myself to accept anything so small and apparently primitive as *Homo.* Either I was wrong on my judgment about Alemayehu's jaws, or else we had in fact discovered *two* species in Hadar, both of which had lived approximately three million years ago. I was pretty sure I knew which alternative was right, but I needed more time to study the collection before committing myself in print.

Even more than time, I desperately needed additional evidence. The Hadar deposits did not disappoint us. The field season the following year yielded up a discovery surpassing even Lucy: a superb trove of almost two hundred hominid fossils from a location called Site 333, including more jaws, leg bones, teeth, foot and hand bones, ribs, vertebrae, cranial fragments, even a part of an infant skull. Oddly, there were very few other animals represented at the site—nothing but hominids littering a steep hillside, and even more when we began systematically to sift through the eroded deposits.

The Site 333 specimens came to be known as "the first family." Whether they were really the remains of a hominid family group is impossible to say, but they were certainly all members of one species. What that gave us was a glimpse through a rarely opened window—a healthy sample of *variation* among individuals of an ancestral population. That is what makes them priceless. It would be very convenient for paleoanthropologists if all the scattered individuals in an extinct species looked just like each other and nothing like the individuals of any other species—but evolution doesn't box life into neat compartments like that. By definition, two closely related species will have a good many traits in common, many of which will show up in the fossil record, while important differences between the species do not. With fossils so scarce and widely distributed, how can one tell which traits are unique to a single species, and which ones are shared?

Compounding this problem is its opposite: Individuals of the *same* species can exhibit a remarkable degree of variation among themselves. A million years from now, if the proverbial anthropologist from Mars comes across the skeleton of a female pygmy in central Africa and then finds another skeleton—this one a male Eskimo—in Alaska, will he know enough to assign both to the species *Homo sapiens*? Maybe not. Even within single populations our species shows marked differences in brain size, facial characteristics, jaw shape, and countless other traits. But if this same alien bone hunter found thirteen perfect skulls together in one place, all of them very much alike, he would probably make the logical assumption that they were members of the same species, even though they were all a little different from each other.

We did not find thirteen perfect skulls at Site 333, but we found the next best thing—fragments of at least thirteen individuals, including nine adults showing considerable variation in size. A cursory study of the new bones led me to suspect that they must belong to the same early *Homo* species as the jaws we'd found the year before. At the end of the season I stopped in Nairobi to show the new collection of fossils to the Leakeys. A number of other scientists were gathered in the National Museum, mostly those working with Richard at Koobi Fora and with Mary at her site of Laetoli in Tanzania. I knew that the fossils would cause a lot of stir. We laid them out on a table and set to talking, everyone picking up bones, rotating them this way and that to catch the light, asking questions, arguing and exclaiming. With the "first family" specimens before them, Richard and Mary were satisfied that *Homo* was indeed present in the Afar three million years ago, though it was obviously a very primitive type. They were convinced, however, that little Lucy was something else entirely.

I left believing they were right—enough at least to present the fossils in print for the first time. In March of 1976 Taieb and I published an article in *Nature* on the Hadar fossils, tentatively suggesting that some of the fossils from the first two seasons showed affinities with *Australopithecus*, while others resembled *Homo*. The 1976–77 field season only produced more evidence to strengthen my conviction. Meanwhile, one of Taieb's colleagues discovered some very primitive stone tools in a gully about three miles from camp. The ability to manufacture implements, no matter how crude, is still taken by most anthropologists to represent the beginnings of human culture. Judging from the potassium-argon dates from a nearby site, these tools were perhaps 2.6 million years old—the most ancient known signature of man's technological ability.

Though the evidence in Ethiopia was still circumstantial, it all pointed to one thing: As long as three million years ago, *Homo* had dwelt in Hadar side by side with *Australopithecus*. The Leakeys had already declared the same to be true at Koobi Fora between three and two million, and at Olduvai between two and one million. All along the Rift Valley, a simple, coherent pattern in time was taking shape. Etched into the center of that pattern—clear, distinct, and very old indeed—was humankind, just as Louis Leakey had predicted. No evidence existed that *Australopithecus* was anything more than the "aberrant offshoot" that the Leakeys had long maintained.

The trouble with simple, coherent patterns is that they tend to obscure messier truths. The evidence from Hadar, Laetoli, and Koobi Fora were about to be marshaled to tell a quite different story, one that would seriously cast doubt upon the Leakeys' familial claim for the antiquity of man. Ironically, the two scientists who would challenge that theory were both present at that meeting in Richard's lab in 1975, when I spread the Hadar specimens around for all to see. One of those scientists was a wiry, long-haired paleoanthropologist named Tim White. I would never have believed it at the time, but the other challenger would be myself.

Fifty miles southwest of Arusha, a dirt road branches off the main route and heads west toward the highlands. As the combi planed over potholes and "washboard" ridges, the landscape grew rockier, the road gouged periodically by old stream beds. Tim peered out of the window, scanning the countryside for exposures. Somebody had told him they had flown over some promising outcrops in the area. It was unlikely that we would pass anything worth investigating right alongside the road, which after all had been used by fossil hunters traveling to and from the Gorge for decades. Tim's vigilance was extreme, and not in the least bit out of character. At the same time he threw questions over his shoulder at Gen, debriefing his student on a cast of an enigmatic skull Gen had examined in Nairobi—the latest hominid emerging from Richard Leakey's dig west of Lake Turkana in Kenya.

After another hour I could just make out the highlands in the haze. Since my first trip to Olduvai in 1972 I had traveled this route many times, but for Gen and Berhane it was all fresh. We spent the time going over surveying strategies, catching up on news, joking nervously about the food we'd be eating for the next several weeks at the Olduvai camp. During field expeditions, Gerry Eck's culinary standards have been known to retreat somewhat from the heights reached by his Chicken Thighs in Orange Cream Sauce. On the scouting expedition the summer before, an endless succession of meals of goat meat and cabbage soup had left us—inexplicably—with a collective longing for peanut butter. There are plenty of peanuts in Tanzania, but I had never seen a single jar of peanut butter. Now I had a surprise for everybody. I reached under my seat and pulled out of my bag a small Waring blender I had packed along with the rest of my gear.

"What the hell is that?" Gen asked.

"Our salvation," I said. "Put peanuts in here, turn it on, and what have you got? Goat antidote."

"You're a clever son of a bitch, Johanson," Tim laughed. "But Mary's got your number. What if she's dismantled the windmills? No windmills, no electricity—no peanut butter. Foiled by the Leakeys again!"

Tim jokes a lot about the Leakeys. It would be safer for everyone to say that it's all meant in good fun, but in fact Tim's battles with Richard and Mary in the past have scarred him deeper than that. Behind his jibes are the memories of the mutual respect and affection he once shared with them both—emotions long since eroded. Tim is too much of a professional to let hard feelings hamper his work. On the contrary, I'd say they push him forward. Whether it stems from his relationship with the Leakeys or from some more distant hurt, there is a driving anger inside him that he harnesses to serve a frighteningly acute, unsparing intellect. I'll admit it makes him difficult to work with for a lot of people. It also makes him a terrific paleoanthropologist, possibly the best. "A damn magician with a bone," as one colleague put it.

When I met him in Richard's lab in 1975, Tim was a promising graduate student: very much a "member of the firm," as those who worked under Leakey's direction were called. Unlike the fractious expeditions that I'd often been associated with, Leakey's field programs were (and still are) models of teamwork. The scientists were bound together by an explicit code of loyalty to the expedition's goals and to those of its leader. This unusual commitment helped to make the Koobi Fora Research Project among the most productive paleoanthropological enterprises in history. But it also tended to blunt expressions of independence. I remembered listening to one of Richard's closest colleagues give a talk at a meeting in London the previous February. The speaker was occasionally prefixing his sentences with the phrase "I believe . . ." Then Richard caught his attention and shoved a piece of paper under his nose. The colleague quickly read the note and then went on with his talk—emphasizing each point thereafter with the phrase "*We* believe . . ." Richard was leaning back in his seat, the edges of a grin showing around his curve-stemmed pipe.

Leakey is known to return such loyalty in kind. He is also a shrewd judge of talent. The year before, Tim had been invited to work under John Harris, the Koobi Fora expedition's paleontologist. Tim's work had already kindled Leakey's interest and respect. He was also appreciated for how well he got along with the "Hominid Gang," the team of expert African fossil hunters that had found most of the Koobi Fora treasures. Whatever bitterness lay ahead in their relationship, in 1975 Tim White was Richard's man. Very soon, however, he would find himself deeply embroiled in the most troubling affair Richard had encountered in his brilliant young career. On the surface, the trouble amounted to nothing more than a disagreement about two different techniques for dating a particular layer of volcanic ash. But resting in the balance was the status of Richard's most treasured find to date: the 1470 skull that had vindicated his father's beliefs and made Richard famous practically overnight.

The terrain of Koobi Fora—or East Rudolf, as it was known in those days—is a baffling moonscape of rocky, gully-pocked undulations, a difficult challenge for any geologist to puzzle out. Volcanic deposits had been hopelessly shuffled about by erosion, as well as by the shifting level of the lake. Ash that had originally fallen elsewhere had also been brought in by rivers and streams that had themselves long since disappeared. Correlating tuffs to make sense of the many square miles embraced by the expedition's three research areas was a daunting prospect.

One of the Koobi Fora geologists was a Harvard graduate student named Kay Behrensmeyer. One day back in 1969, Behrensmeyer had discovered some crude stone tools in a grayish layer of volcanic ash, very much like the early artifacts found at Olduvai Gorge. This discovery was important for two reasons. Richard had already found his first *Australopithecus boisei* skull in another area to the north, in what was believed to be the same "horizon," or geological slice of time. Australopithecines were not thought capable of manufacturing even the simplest stone tools; hence the artifacts in the ash were the first hint that some species of *Homo* might have inhabited the site. Adding to the intrigue was the nature of the ash itself: The layer contained exactly the kind of crystals that lent themselves to dating by the potassium-argon method. Here then was an opportunity to assign a date to the tools embedded in the ash. Anything subsequently found in the deposits *below* the "Kay Behrensmeyer Site"—or KBS for short—would have to be older still.

Richard quickly sent off samples of the crystals to two British geochronologists named Frank Fitch and Jack Miller. Miller had run potassium-argon dates for Richard's father at Olduvai Gorge, but since that time the pair had been instrumental in developing considerably more sophisticated techniques. Their first response came back with a date for the tuff of 221 million years: clearly a false reading, unless one wants to believe that dinosaurs made stone tools. It turned out that the deposits from which Leakey and Behrensmeyer had collected the samples were contaminated by much older volcanic material brought down from higher elevations by rivers and streams.

Leakey went back to the KBS Tuff and collected what he hoped were purer samples and sent them off to Fitch and Miller. This time they came up with more gratifying results. The crystals were first dated at 2.4 million years; a further test pushed the date back to 2.6 million. That meant that the tools Behrensmeyer had found were the oldest indication of human habitation on earth, almost a million years older than anything found at Olduvai Gorge. Three years later, along came 1470—found well *below* the KBS Tuff. That was how Richard could announce with confidence that the new skull was at least 2.6 million years old, and probably closer to 3 million. With his "Oldest Man" securely in hand, funding for the Koobi Fora Research Project poured in. More fossils were found in 1973, including dozens of new hominid specimens. Meanwhile, geologist Ian Findlater completed a stratigraphic map correlating the complex volcanic tuffs at the three Koobi Fora research areas into a simplified, workable scheme.

Things were looking very sunny in Leakey's camp. Only one nagging doubt still needed to be put to rest, a little cloud on the horizon that took the shape of a pig. At first the threat it embodied could easily be shrugged off. But before Richard was through on the eastern shore of Lake Turkana, the shadow of the pig would spread to cast a pall over the entire enterprise.

As part of his site development, Leakey had very early on invited palentologist Basil Cooke to conduct an analysis of the rich collection of suids—members of the pig family—emerging from Koobi Fora. Cooke was the biostratigrapher who had used the evolving changes in extinct pigs to develop the Omo's "yardstick" of evolutionary time. He had also worked with Louis Leakey to develop a similar time scheme for the pigs of Olduvai Gorge. If the Koobi Fora suids could be sorted out and compared to those from the Omo and Olduvai sites, Richard thought, the biostratigraphy would bolster confidence in the radiometric date given by Fitch and Miller for the KBS Tuff. Simultaneously studies were under way with yet another dating system, one that uses regular, periodic reversals in the earth's magnetic field to determine the age of rocks that can be magnetized. This paleomagnetic study was producing a reassuring match with the 2.6-million-year potassium-argon date. Confirmation with biostratigraphic evidence would all but clinch the matter.

Unfortunately, Cooke's study did just the opposite. After months spent trying to make sense of the suids, he concluded that they indicated the 2.6 radiometric date was wrong. The fossils Cooke examined from the KBS Tuff matched instead the pigs in the Omo from much farther up in the succession, closer to 2 million years. Leakey and his colleagues were disappointed, perhaps, but not too bothered by Cooke's initial results. There had been some mistakes in the collection techniques early on, and who knows, maybe the real error was to be found in the Omo and Olduvai dates. They adopted a wait-and-see attitude. In the meantime, Richard stuck firm behind the radiometric study, declaring when 1470 was announced that the KBS Tuff had been "securely dated" at 2.6 million years.

Outside the Leakey camp, however, Cooke's pigs raised a few eyebrows—especially across the river in the Omo, only fifty miles north from Koobi Fora. Earlier, Clark Howell had expressed some skepticism about the Fitch-Miller radiometric date. For one thing, the doubtful pigs were supported by biostratigraphic studies on extinct elephants, antelopes, and other mammals. For another, Fitch and Miller's own study had never seemed rock solid to begin with. Two out of three of the reviewers of a subsequent paper had recommended that it be rejected; one commented that it "did not even approach the standards required of a scientific journal." That paper defending the 2.6-million-year date for the KBS Tuff was published nevertheless, virtually unchanged. Finally, results of an independent potassium-argon analysis conducted on KBS crystals by Garniss Curtis were indicating pig-ish dates of 1.8 and 1.6 million years.

By 1975 clear sides had formed on the issue. The Koobi Fora camp—mostly British scientists in Nairobi—rallied around the 2.6-

If enough is known about a species' lineage, the appearance and disappearance of particular forms can be used as a measure of the passage of time. In this illustration, changes in the chewing surface of the teeth of the pig genus Metridiochoerus *are placed in the context of potassium-argon (K/Ar) dates. The teeth can then be used to help determine the age of fossil deposits that may be undatable by the potassium-argon technique.* ILLUSTRATION BY DOUGLAS BECKNER. AFTER PHOTOS FROM T. D. WHITE AND J. M. HARRIS, *SCIENCE*, OCTOBER 7, 1977.

million-year KBS Tuff date reached by the apparently objective sophistication of Fitch and Miller's radiometric analysis. Howell's Omo group, comprised primarily of American scientists working out of Berkeley, threw their weight behind Basil Cooke's 2.0-million-year date, arrived at through tried-and-true biostratigraphic analysis. This tug-of-war was not some idle academic recreation: The rope stretched between the two camps was the human family line itself. If Cooke was right, then Leakey's 1470 was really no older than the *Homo habilis* specimens at Olduvai Gorge, and his argument for dismissing *Australopithecus* from the human lineage would fall apart.

In spite of its seriousness, the KBS controversy generated some of the great light moments in modern paleoanthropology. At one meeting held in an old castle in Austria, two scientists on opposite sides of the issue suddenly grabbed swords hanging on the castle walls, leaped up on a table and began to duel. (They had planned and practiced the melodrama the night before.) Koobi Fora archeologist Glynn Isaac once showed up at another conference wearing a "pig proof helmet." At a subsequent meeting in London in February 1975, Basil Cooke appeared sporting a novelty necktie with the letters MCP woven into it, standing for "Male Chauvinist Pig." By this time Cooke had amassed much better data on the Koobi Fora suids, particularly on a species called *Mesochoerus* that lived throughout East Africa about two million years ago; everywhere, that is, except Koobi Fora, where—if you believed the Fitch-Miller date for the KBS Tuff—it inexplicably appeared over half a million years earlier. At one point in the meeting Cooke stood up and pointed to his necktie. "You might think you know what MCP stand for, but you don't," he said. "It really stands for '*Mesochoerus* correlates properly.' "

Leakey was not amused.

"I was prickly and upset," he recalled later.* "I felt that Basil, as a member of 'my' team, should not have used the Koobi Fora data in the way he did without giving me a full report before the meeting." Cooke was mystified by this; his views on the pigs had not changed since the last time he had aired them, so he had seen no reason to solicit Richard's approval before presenting his work.

Equally mystifying to those of us who attended that London meeting was the "ecological hypothesis" that Leakey, Isaac, and John Harris were proposing to account for the 600,000-year difference in the faunal record between their site and the Omo. Their idea was that in environments separated by a geographic barrier—in this case, the wide Omo River—populations of animals might have evolved at different rates. The precocious pigs of Koobi Fora therefore only *appeared* to be half a million years younger than their counterparts across the river, because they had been evolving at a faster rate. Charles Darwin himself had deduced from his studies of finches in the Galápagos Islands that populations isolated from one another would take unique evolutionary paths. Most evolutionists believe, in fact, that geographical isolation is necessary for new species to evolve. But the geographical barriers Darwin was talking about were ocean expanses, not rivers. More important, such barriers would affect the *direction* of change, not the pace. It seemed to me that Richard's people were stretching a bit to justify their belief in the Fitch-Miller date. Elephants and other migrating browsers cross the Omo occasionally even today. It's hard to imagine them meeting their own evolutionary grandfathers on the other side.

If few people were impressed with the ecological hypothesis,

---

*As reported in Roger Lewin's book *Bones of Contention*, which contains an excellent, detailed account of the KBS Tuff controversy.

Leakey still had good reasons to remain committed to his 2.6-million-year date. Preliminary results on the KBS Tuff using another radiometric technique, called fission-track dating, were coming in very close to the potassium-argon results. By this time Richard also had an inkling that Basil Cooke's pig dates were questionable—thanks in part to Tim White. After collecting various fauna with John Harris in the field, Tim had gone back to Nairobi the year before and photographed some of the pigs previously collected by Cooke. He hadn't done any analysis yet, but what Tim and John had seen caused them to wonder out loud to Richard about Cooke's sorting methods. Perhaps further fieldwork would resolve the discrepancies.

Meanwhile, Richard still had another card up his sleeve at the London conference. He played it at a small party he gave after dinner one evening. Glynn Isaac was there, along with Kay Behrensmeyer and physical anthropologists Bernard Wood and Michael Day—all staunch "members of the firm" at Koobi Fora. But Richard had also invited Clark Howell, geologist Frank Brown, and some of their colleagues on the other side of the issue. The talk kept coming back to the KBS date, the pigs, the crystals, the geology, eventually descending into a shouting match over who was right and who was wrong. Richard just sat there, listening. Suddenly he got up, walked to another room and returned with a cast of a beautiful hominid hipbone—a big thing, *Homo erectus* by the look of it. Everybody was quiet.

"Here's a little something one of my Hominid Gang picked up during the off-season," he said. "It came from below the Tulu Bor Tuff at Koobi Fora." He settled back in his chair and smiled. We all knew that the Tulu Bor Tuff was *below* the KBS Tuff. Meaning that this amazingly humanlike pelvis was at least three million years old.

"That's impossible," Howell said.

"There it is," Richard replied, sucking on his pipe. "Let the fossil speak for itself." That started up a whole new argument.

Leakey was sure enough about the Tulu Bor pelvis to write a draft of a paper on the find for *Nature*. But what he'd heard from Howell and others convinced him that he'd better make doubly sure. In the summer of 1975 he sent Harris, White, and the Hominid Gang out to Koobi Fora alone to verify the provenance of the fossil—in other words, to check that the pelvis had been placed in the proper geological and paleontological context.

"We went out to the place where the pelvis had come from and did some collecting," Tim remembers. "The pigs we found were nothing like the ones below Tulu Bor—it wasn't the same tuff at all. That was the first hint we had that the Koobi Fora stratigraphy was all screwed up."

Quickly Tim and Harris reported back that the pelvis had really come from much higher up in the succession, and could not be older than 1.9 million years. The *Nature* paper was withdrawn, saving Richard considerable embarrassment. Ian Findlater was called down from London, where he was working on his dissertion, to take another look at the strata. He admitted there was the possibility of a mistake.

If the news on Tulu Bor was bad, Richard still had high hopes for what Harris and White would find out about the pigs. Basil Cooke was still expecting to spend the 1975 field season at Koobi Fora and continue his own analysis of the suids. Leakey discouraged him from coming. Instead he urged Harris and White to write their own monograph on the suids of Koobi Fora.

"The chronology that Basil had put together relied on only one evolutionary line of pigs, which is what he'd based his time correlations on between Koobi Fora and the other East African sites," Tim later told me. "But John and I were having trouble making sense of that scheme. We saw evidence of change in three separate pig lines, not one. When Richard heard that, he thought, 'Hey, if Cooke's pig family tree is screwed up, probably his dates are too.' That's what he wanted us to prove in the field. So he could keep his ancient *Homo* alive. But you can't always get what you want."

By the end of the summer the two young paleontologists had gathered enough new material to know that Leakey's 2.6-million-year date for the KBS Tuff was in trouble. Cooke's time correlations looked prerry good after all. There was still one slim hope: Though the sub-KBS pigs were correlating very closely with the Omo pigs dated at 2.0 million, perhaps the *Omo* date was misleading. To follow that angle, Tim would have to wait until early the following year, when he could return to the States and study the part of the Omo paleontological collection housed in Berkeley. What Tim found there only strengthened his suspicions. At the end of March 1976, Richard flew out to the West Coast himself to give a seminar. Tim picked him up at the Oakland airport.

"On the way back I told him everything," he remembers. "We went over the metrics, the geology, the stratigraphy, the whole damn analysis. The only thing he was displeased with was the dating on my chart. He was worried about the conflict with the potassium-argon results."

The last step in the analysis was to take another look at the rest of the Omo pigs, stored in a laboratory outside Paris. Harris joined Tim there in May, and after a week of study they had eliminated all doubt. They flew on to Nairobi in June and gave Richard the news. He offered no comment. John and Tim set to work writing up their monograph.

It looked to be a blockbuster. First, they would greatly expand on Cooke's original biostratigraphic work on the Plio-Pleistocene suids. Included for the first time were the species found in South Africa: an extremely important addition, since in the absence of the kind of volcanic material needed for potassium-argon dating, biostratigraphy remained the only reliable means of estimating the ages of the South African hominids found by Dart and Broom. Second, the pig data they presented would by its very nature trigger a complete reappraisal of the Koobi Fora geology and the tuff-numbering system in use there. Third, the pigs would prove that the radiometric date of 2.6 million for the KBS Tuff was skewed, for whatever reason. With the fall of that date, Leakey's Oldest Man would topple too.

"From the standpoint of the science, that last point was by far the least important," Tim says. "But that's what got everyone bent out of shape."

Richard had heard enough from different points of reference to know that the stratigraphy of his site was in serious trouble. In August he called a "summit meeting" in the field of everyone involved, including Glynn Isaac, Kay Behrensmeyer, and Meave. Ian Findlater flew down again from London. John Harris and Tim came up from Nairobi, bringing a draft of their pig monograph with them. The group drove from site to site, John and Tim pointing out the problems they saw with the tuff correlations on paleontological grounds, while Ian and Kay defended their original geological interpretations. After several such outings everybody could see that what had once seemed like a well-integrated scheme was now full of holes. Then Isaac proposed a temporary resolution: Instead of attempting to trace the volcanic tuffs *between* the research areas, each of the three areas would be mapped out with its own complex tuff-numbering system. Under Isaac's scheme, it would be very cumbersome, if not impossible, to pin *any* date of the fossils found between the tuffs. It was a compromise that pleased no one, but John and Tim reluctantly agreed to go along. They let everyone know they were planning to abstract from their monograph an article for *Nature* on the implications of their pig work. Isaac began drafting a manuscript to accompany it, explaining the new tuff-numbering system.

If Richard was upset by the direction in which his two paleontologists were leading him, he didn't show it. Meanwhile, there was better news from other fronts of the battle. Tony Hurford and Andrew Gleadow, the two geochronologists who were pursuing a fission-track date for the KBS Tuff, had refined their technique and could confidently report the KBS Tuff clocked in at 2.44 million years, give or take 80,000 years. This was a little younger than the Fitch-Miller date of 2.6 million, but even that trifling discrepancy quickly melted away: In the same issue of *Nature* that presented the fission-track study, Fitch and Miller revised their date upward to 2.42 million years. Some colleagues found this seamless agreement between the two radiometric methods almost too good to be true, but it certainly bolstered Leakey's position. Add to that the revelation that Garniss Curtis's reading of 1.6 million years had proved to be flawed by an improperly adjusted weighing balance, and it is perhaps no wonder Richard remained loyal to his geochronologists, pigs or no pigs.

Tim and John Harris finished their own paper to *Nature* early in September, including a discussion on the implication of the work for the hominids. Both men are convinced that Leakey, Isaac, and the other members of the Koobi Fora team were well aware of the conclusion they would present: The KBS Tuff was 2 million years old, not 2.6, and the famous 1470 skull and other hominids found below it were also much younger than claimed. John gave the draft to Leakey in his office in Nairobi. On September 7 Tim had a hurried meeting with Richard, who told him he didn't agree with the conclusions of the paper, but gave no indication that he was seriously displeased. The same day, Tim was scheduled to leave to attend an international meeting in Nice. But before leaving the museum, he stopped in at John Harris's office. What he learned there appalled him.

"John told me that Leakey was furious about our *Nature* paper," Tim recalls. "He had complained to John that we had not consulted him properly, which was nonsense—we'd spent the last two weeks of August outlining our intentions to everybody at Koobi Fora. Richard objected to our referring in the paper to hominid fossils that he claimed had not been published. Supposedly this was against a long-standing policy in the East Rudolf Research Project. I lost my temper over that. *All* the hominids we cited had been published by Richard himself in *Nature* and *The American Scientist.* He just didn't want us to present our evidence because it wasn't what he wanted to hear."

Reportedly, Leakey himself remembers the incident quite differently. Roger Lewin quotes him as saying that he told Tim about his objections directly, reminding him that the group's policy forbade any group member to mention fossils in print until the fossils had been formally described in *The American Journal of Physical Anthropology.* The rule, Richard explained, was meant to protect whoever was doing the nuts-and-bolts work on the material from being internally scooped by other members of the team. According to Leakey, Tim then exploded in rage, announced his resignation from the Koobi Fora project and stormed out of the office, slamming the door.

"That makes an interesting story, but it never happened," Tim tells me, and his journal entries from the time bear him out. "Furthermore, we didn't say anything about the hominids in that manuscript that a reader of *Nature* couldn't have said. We followed group protocol by giving Leakey a presubmission copy of our paper. It was insulting to be told secondhand that this was somehow wrong."

Badly shaken up, Tim left for the airport. If the *Nature* paper was to be emasculated by removing all reference to the implications for the hominids, then Tim wanted no part of it. John Harris, a better diplomat than his co-author (as well as Richard Leakey's brother-in-law), prevailed upon Tim to allow the paper to be rewritten with the implications for dating couched in ambiguous, less offensive wording. Harris then submitted the manuscript to *Nature* as planned—only to have it rejected. The editors claimed that it did not add significantly to what Basil Cooke had already published about the pigs.

Tim scoffs at that explanation, and it is in fact hard to justify. Whatever else the manuscript had in its favor, it revealed the time correlations for the South African pigs—information that had never been published and that would later prove absolutely essential to the reconstruction of the human family tree brought about by my discoveries in Ethiopia. Tim still suspects the *Nature* editors or reviewers had been influenced to reject the manuscript, but there is no way of proving that. Meanwhile, he and Harris submitted it unchanged to *Science,* where it appeared in October 1977. Thereafter, Cooke brought his own pig phylogeny into line with theirs. Tim made an attempt at patching things up with Richard, but their relationship cooled, eventually reaching absolute zero.

"You have to understand that the 2.6-million-year date was like a religion at Koobi Fora," Tim remembers. "My research reached a different conclusion, and Richard did his best to suppress it. No personal friendship, no political consideration, makes that acceptable to me. I was revolted by the whole thing."

As things turned out, the Fitch-Miller date for the KBS Tuff was proved to have been in error after all. A study conducted by a third, impartial geochronologist confirmed Garniss Curtis's date of 1.8 million. Further fission-track studies by Hurford and Gleadow came up with the same number. Richard abandoned his defense of the older date. The whole bitter episode had taxed his health, which was already seriously threatened by kidney failure. In 1979 a kidney transplant from his brother Philip was required to save his life.

Meanwhile, Tim White went on to work on the remarkable collection of hominid jaws that Mary Leakey had found at Laetoli, south of Olduvai. If Tim had played the devil's advocate at Koobi Fora, at Laetoli he'd prove to be the devil himself.

CHAPTER FOUR: *Breaking Ranks*

It's not just what we inherit from our mothers and fathers that haunts us. It's all kinds of old defunct theories, all sorts of old defunct beliefs, and things like that. It's not that they actually live on *in us; they are simply lodged there, and we cannot get rid of them.*

—Henrik Ibsen, *Ghosts*

ON THE WAY TO OLDUVAI, shortly before the road begins to wind up the steep ascent to Ngorongoro Crater, there is a village called Mto Wa Mbu, which in English means "River of the Mosquitoes." It is a busy little town, with cattle, goats, and chickens in the streets and a constant traffic of buses and trucks. Mto Wa Mbu was our last opportunity to stock up on fresh fruit and vegetables.

As we fanned out through the network of stalls in the market, we soon found ourselves tailed by a group of young boys. Much to Gen Suwa's embarrassment, the boys decided to pretend he was Bruce Lee. Wherever poor Gen turned he was confronted by a flurry of martial arts gestures and unbridled giggling. Tim's wide-brimmed hat with its rattlesnake band was the next most popular attraction. Tim was comfortable with the kids. He would pretend not to notice them, and then suddenly he would whip around and chase them, waving his arms and charging into the group with a snarling grin. This display brought almost as many squeals of delight as the spectacle of a Kung Fu warrior wearing thick glasses and a blue UC Berkeley windbreaker.

The market at Mto Wa Mbu, on the road to Olduvai Gorge DONALD JOHANSON

We bought some oranges, green bananas, jackfruit, cabbage, and tomatoes. Back in the combi, we cut open a jackfruit for lunch and passed pieces around stripping the delicious, fragrant fruit off its skin with our teeth. Pretty soon the road began to climb the escarpment above Lake Manyara National Park. Near the top we pulled off to the side of the road and walked over to the edge of the cliff. In the valley below, Lake Manyara looked like a silver plate set out on a mottled green cloth. The green was darkest along the west shore of the lake, with another swatch of forest lining both sides of the snaking river that fed the lake from the north. On the far side, a cooler jade outlined the salt flats, infused in one place with a faint pinkish blush. While I watched, a ripple moved through the pink patch. Momentarily the color dispersed, then coalesced in another place. It suddenly dawned on me that something down there had startled a flock of flamingos. On our side of the lake, I could just make out a family of elephants browsing in a gallery of fever trees, flickering in and out of view like stones glimpsed through the water of a running brook. An eagle slowly circled above them, riding the updraft against the escarpment.

"Gen, Berhane—take a look at this," Tim said, pointing down into the valley. "Freshwater streams running into a wide, shallow alkaline lake. Gallery forests, marsh flats. This is the best modern analog for what Olduvai was like when Zinj was running around about two million years ago."

Good point. In the early Pleistocene, Olduvai wasn't a gorge at all, but rather an open plain surrounding just such a lake as this. It would be a busy spot, a place flickering with elephants and a menagerie of other herbivores attracted by the swampy richness; a place, like Manyara today, populated as well by lions, hyenas, and other carnivores. They all left their bones by the Olduvai lake millions of years before there was such a thing as the Serengeti Plain. Hominids too. We looked down in silence. As if we were all thinking the same thing, the four of us turned and walked quickly back to the van. If we pressed on we could still make the Gorge in time for a little poking around in its famous gullies before dinner.

Back in 1975, when Tim was still a valued "member of the firm," Richard Leakey had recommended him for a tremendously important assignment. Under Mary Leakey's direction, research expeditions to the Tanzanian site of Laetoli near Olduvai Gorge had collected some promising new hominid jaws and teeth from deposits believed to be over three and a half million years in age—older even than my Hadar finds. In all there were thirteen specimens in the collection, very primitive, but also showing some curiously *Homo*-like features. The best of the lot was a well-preserved mandible with nine teeth in place, labeled L.H.-4. Tim was given the job of describing the jaws for publication. He was twenty-five years old at the time and still two years shy of his Ph.D. To be invited to formally describe the hominids of Laetoli was a fantastic opportunity.

Tim was well into the work when I met him that day in Richard's lab. I had learned a little bit about the Laetoli hominids from Mary, and Tim had of course heard about Lucy and the other fossils emerging from Hadar. He had also heard something about me.

"Richard had warned me about this hotshot up in the Afar who didn't know what he was doing," Tim joked to me months later. "But I was waiting to make up my own mind. You turned out to be not half as dumb as Richard made out."

If Richard had any real doubts about my expertise he certainly was hiding it well. And I was placing great faith in his. I was all but convinced that we had discovered at least two species in the Afar: The larger jaws represented a very primitive species of *Homo*, while little Lucy and her ilk were something quite different, probably a primitive, though gracile (slenderly built), australopithecine. This arrangement was of course very much to Richard and Mary's liking. In those years they agreed on very little, but both were equally—and utterly—committed to Louis's faith in ancient *Homo*. I was taking my cues from them—not to curry favor, though being friends with Richard was unquestionably an advantage for any young paleoanthropologist working in East Africa—but because the Leakeys' scheme seemed the best way to accommodate the perplexing variation evident throughout the Hadar collection. I was also very much aware of the kind of attention that would follow anyone who spoke the words "I have found the oldest man." The hubbub around the Hadar fossils strewn about that lab table was just the beginning.

Tim's first words to me went right to the point. When the others had left the room, he emerged from a corner and approached the fossils. After he'd examined them for a few minutes he turned to me and said, "I think your fossils from Hadar and Mary's fossils from Laetoli may be the same." That startled me. Laetoli and Hadar were more than a thousand miles apart, and their hominids separated in time by at least half a million years. It seemed very unlikely. Then Tim showed me a couple for comparison. They were nearly identical.

"One species?" I said.

"One species."

"What about Lucy?

"Lucy too."

"No way," I said. "She's much too small and primitive. Jaw shape, teeth, brain size—none of her makes sense with the others." I mentioned half a dozen other primitive traits. "Come to Cleveland and I'll show you."

"Lucy too."

"Come to Cleveland."

I wasn't buying Tim's idea, but his forthrightness and his obvious command of hominid morphology impressed me. We agreed to keep each other up to date on our work.

In March of 1976 my *Nature* article with Maurice Taieb appeared as planned, strongly hinting that there were at least two species represented in the Hadar, and possibly three: gracile and robust forms of *Australopithecus*, side by side with primitive *Homo*. I would meet Tim occasionally at meetings over the next year, and he would rarely fail to bring the subject up again. I listened with interest, but having committed myself in print, I was more inclined to find reasons to dismiss new points of view than I was to reexamine my own conclusions.

I didn't have to look far for an explanation for Tim's assessment. As a student of Milford Wolpoff and Loring Brace at the

University of Michigan, he had been trained in a tradition of "super-lumpers." By emphasizing the similarities among the various known hominid specimens over their apparent differences, the Michigan school had once trimmed the human family tree down to a simple trunk: At any time in the past, they maintained, there was only one kind of hominid living, just as there is only one today. The discoveries of the 1970s had rendered the "single species hypothesis" virtually indefensible; even Loring Brace had difficulty placing a relatively brainy skull like 1470 in the same species with the robust australopithecines, with brains the size of grapefruits, found in the same time period at the same site. But Brace and his colleagues at Michigan still looked unsparingly at any attempt to create rigid species distinctions on the basis of what they thought of as minor morphological differences.

"Of course Tim White sees only one species at Hadar," I thought to myself. "His eyes are trained to look for the sameness among things." I didn't know at the time that he had already quarreled with his teachers.

Even so, Tim's insinuations made me uneasy. At the time, I was struggling with a burden that most paleoanthropologists gladly would have shared: I was finding hominid fossils at a much faster rate than I could analyze them. Had I published too soon after all? I was almost relieved when the revolution in Ethiopia put off hope of a field season for 1977—at last I would have a chance to catch my breath and sit down to evaluate the collection in earnest. I had no inkling, of course, that I would not be allowed to return to the Hadar for many years to come.

Tim too found himself with some time on his hands in the summer of 1977. After the KBS dating debacle blew up in his face, he had not been asked back to work at Koobi Fora. The previous fall he had instead returned to Nairobi to continue work on the Laetoli descriptions and finish his thesis. He spent a lot of time working with Mary Leakey. Tim was alone when Christmas rolled around, and Mary invited him to celebrate the holidays at the Olduvai camp. She seemed to take to him, and Tim returned her friendship in kind.

"I felt quite a bit of sympathy for her," he remembers. "It went beyond the image of this tough, crusty lady living out in the middle of nowhere. Her husband had died. He hadn't been such a great husband in the first place. Her sons Richard and Philip hated each other. But she stuck with it, down there at Olduvai year after year with just her animals. I had a lot of respect for that dedication."

His work with Mary concluded, Tim came back to the States and visited me in Cleveland, bringing casts of the Laetoli fossils with him. We were anxious to have another look at them side by side with the much more extensive Hadar collection. We laid the fossils out on the counter in my lab and went to work. One thing was abundantly clear: The larger jaws from Hadar looked astonishingly like those from Laetoli. Knowing that superficial resemblances can be misleading, we worked over the specimens for days, one jaw, one palate, one tooth at a time, measuring distances and angles and examining the morphology of every cusp, trying to prove that our first impression was wrong. It wasn't. The teeth and jaws were virtually identical; it would be hard to find a single morphologist in the world who would dispute that. But I doubt that you could find two who would readily agree on what exactly they were.

Originally, both of us were inclined to see the big jaws as *Homo.* They did not have the large molars and heavy buttressing characteristic of australopithecine jaws. But then they also exhibited a puzzling array of very primitive, almost apelike features. Their arcade of teeth, for instance, resembled the boxlike ape pattern rather than the parabolic shape characteristic of hominids. The more we analyzed the teeth and jaws, the more they seemed to occupy some intermediate zone between apes and all other known hominids. But if they were *Homo,* how could they be more primitive than their own australopithecine ancestors?

And then there was Lucy. Her jaw was very different. First of all it was clearly smaller, Second, the dental arcade was V-shaped. What pieces we had recovered of her cranium, meanwhile, showed that Lucy's brain was much too little to qualify as *Homo* by any conventional standard; as a matter of fact, it wasn't much bigger than that of a chimp. Halfway through the analysis I was more convinced than ever that she was a different species from the bigger jaws at Hadar, including some from the First Family site. On the other hand, some of the First Family jaws were small and Lucy-like.

"How are you going to fit *that* problem into your notion of two species?" Tim taunted. It was 3:00 A.M. and we'd been working and arguing since the previous morning. "Or do you have two species represented in one family?"

"Don't be silly."

"It's the logical conclusion of *your* argument. Not mine."

"So how would *you* explain it?"

"I told you before. Sexual dimorphism. Combined with allometry."

Tim had been pushing these Michigan-influenced arguments from the beginning to explain Lucy's anomalies. Sexual dimorphism refers to the general morphological differences apparent between the two sexes of a species. These can include dissimilarities in canine-tooth size, cranial capacity, pelvic measurements, limb-bone dimensions, and a host of other features. Body size is a common example. The males of a pronounced sexually dimorphic species like the gorilla outweigh the females by as much as two to one. The sexes of less sexually dimorphic primates, including ourselves, are more similarly proportioned.

Allometry is a related phenomenon. Basically, it means that differences in anatomical *size* will lead to predictable changes in *shape* as well. You cannot "grow" a male gorilla just by expanding the dimensions of a female gorilla's anatomy. His canines are not just larger than hers, they are *proportionately* larger as well, and their shape is different. Similarly, a female amounts to more than a scaled-down model of the male. This was the argument that Tim was using to explain Lucy's oddly shaped jaw. Her front teeth were not only smaller than those of her male counterparts—they were proportionately smaller too. Since her jaw did not have to accommodate fuller front teeth, it could narrow toward the front into that distinctly V shape.

Allometry or not, I still could not accept Tim's argument. Lucy's jaw was simply *too* different from the rest. This kept leading me back to her other points of difference—her small stature for instance—even though I'd already conceded to Tim that Lucy's size could be accommodated with the others in a sexually dimorphic species. In the back of my mind too was that little brain. I kept thinking how neatly one of my larger Hadar mandibles would fit on to Richard's 1470 skull, so much more advanced than Lucy's, and what *good sense* that would make out of the whole human evolutionary puzzle. Primitive *Homo*, walking erect three million years ago, brain enlarging—beautiful. I didn't need Tim to tell me there was little substance to this argument: From what we had recovered of their crania, we had no reason to believe that the larger Hadar hominids had brains any bigger than Lucy's. So I didn't talk about brain size—I merely *thought* it. What I talked about was jaw size and shape.

"Sure, Lucy's jaw is different, if you look at her up against the other extreme," Tim said to me one day. To demonstrate he placed Lucy's jaw on the counter next to the largest one from the First Family site. "But take your Leakey lenses off for a minute and look at *this*."

Putting those two fossils on either end of the long counter, Tim walked over to the shelf where we kept the other jaws in the collection. One by one he laid them down between the two extremes, so they formed a graduated series from largest to smallest. Suddenly Lucy didn't look so strange after all. The difference in size was quantitative, not qualitative; up against the jaws in the middle, for instance, Lucy's jaw didn't seem so unusual.

"All right, I'll concede that we can't argue for two species on the basis of size. But what about the jaw shape? You say allometry. If you can prove that, you win."

"You're on."

That took some time and a lot of patience. But when we had finished scaling down all the jaws in the collection to Lucy's size, the differences in shape all but disappeared. The last obstacle to Tim's claim for one species had been removed. The analysis had taken all summer and covered a lot more technical ground than I've described here. But in the end, I had to concede that all the fossils of Hadar and Laetoli comprised only one species. The next challenge was to figure out what it was.

Tim left at the end of the summer and I settled down to more routine tasks at the museum. Having worked around to accepting the plausibility of Tim's hypothesis, I gradually began to see how well it accounted for what I had once seen as unresolvable ambiguities. Topping the list was that vein of primitiveness running through the Hadar and Laetoli collections, combined with what were unmistakably *Homo*-like traits. I had been arguing from the preconception that *Homo* equaled advanced. Tim's line of reasoning demanded a mental flip-flop: Look on the *Homo* pattern as the primitive one, with the big-molared australopithecine pattern arising *later*. My two-species bias had pushed me to draw clear lines between categories that simply did not apply: human versus australopithecine, gracile versus robust, advanced versus primitive. Only after I'd pried myself free from the safety and convenience of those distinctions could I see what we *really* had: a single, very old, very primitive hominid species that was ancestral to *both Homo* and *Australopithecus*. The implica-

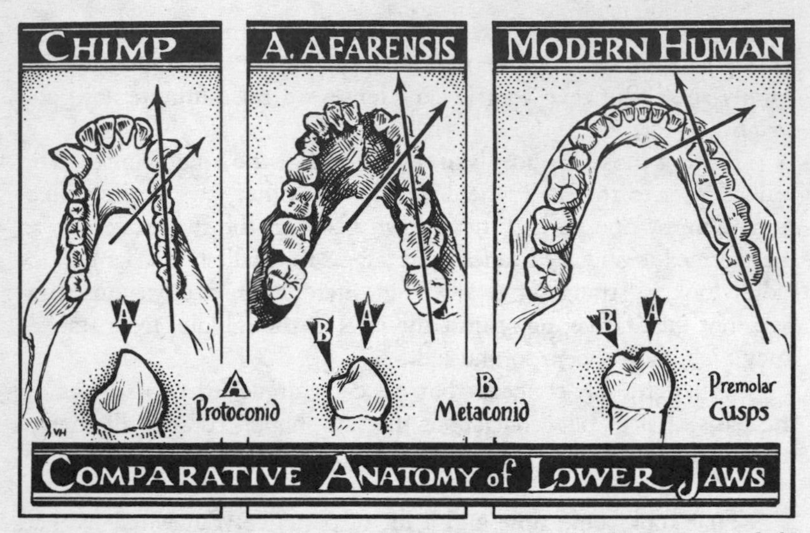

The primitive characteristics of jaws and teeth led Johanson and White to conclude that A. afarensis *was the ancestor of all later hominids. This illustration shows that the premolar of* afarensis *no longer has the cutting edge of a chimpanzee's premolar, but neither has it yet evolved into the true two-cusped form present in later hominids. Its crown is also set at a sharper angle to the tooth row, like a chimp's.*

tions of that were flabbergasting. What we had in hand was the common ancestor to all other known hominids.

In November I wrote to Mary Leakey and told her that contrary to what I'd already published, I was now almost certain that my Hadar finds and her Laetoli ones represented a single species. ("Johanson's been '*white*washed!' " she joked to a colleague.) Though Tim and I had undertaken the interpretive analysis of the fossils, we considered that Mary, as the finder of the Laetoli hominids, fully deserved to be one of the authors of the paper we would eventually publish. (According to the terms of a formal agreement of the Afar Research Expedition, Yves Coppens, the director of the French team in the Afar, had the option to co-author the paper as well.) In my letter to Mary I suggested that we consider proposing a new *Homo* species, more primitive even than *Homo habilis.* She replied that it was too early to assign a species name—better to wait until we had some better cranial material in hand. But that reasoning could hang us up forever. Meanwhile, I suspected that Coppens was planning to use the Hadar specimens soon to name his own new species. After the summer spent day and night with those bones, I felt that Tim and I knew their import better than Yves, Mary, or anyone else. We couldn't waste time waiting for skulls to push out of the ground. We already had plenty to go on.

Early in December Tim returned to Cleveland for another, we hoped final, confrontation. We were very much aware that whatever we decided to call the species would raise a storm. Proposing a new species means much more than sticking a label on some bones so everybody knows what to call them. The creation of a taxon is a formal recognition that this creature *existed*—distinct, viable, reproducing with only its own kind, embodying an evolutionary past and future all its own. A species name defines its relation to all previously known species in its family tree. Unless the new evidence exactly matches what could have been predicted from the old, it will cause some changes in the tree that are bound to upset people. We were young, and had done nothing yet that would prompt unqualified respect. By opting for a single species at Hadar, I was in effect conceding that I had already made one serious mistake. Tim was developing a reputation as an able morphologist, but his brittleness wasn't making him many friends along the way. Why should anyone believe us? We had to be sure there was no reasonable argument against us that we hadn't already thought of ourselves.

The first thing we did was place the new, still unnamed species in a chronological context. The ages of most of the East African hominids were pretty well known through potassium-argon dating. To date Leakey's 1470 skull, the first known evidence of *Homo,* we used Garniss Curtis's radiometric date of 1.8 million for the KBS Tuff above it in the sequence. The hyper-robust Zinj (*Australopithecus boisei*) was contemporary with *Homo habilis* at Koobi Fora and Olduvai, so these two lines were clearly separate. The robust australopithecines disappeared from the fossil record just before one million years ago. *Habilis,* meanwhile, continued on to *Homo erectus* and eventually *Homo sapiens.* We drew a tree with two branches, the *Homo* line and the robust australopithecine line, with the latter branch lopped off by extinction at 1 million years. At the base we place the unnamed Hadar Laetoli hominid, dated between 3 and 3.5 million years old.

That left only the gracile *A. africanus* form to fit in. For the South African australopithecines, we relied on an age estimate between 2.7 and 2.1 million years, developed through biostratigraphic studies by Elisabeth Vrba, a South African paleontologist, as well as through Tim's correlations from his African pig study with John Harris. Raymond Dart had believed his species was ancestral to *Homo*—so did everyone else who believed that the human line arose from an australopithecine. But that no longer seemed to make such good sense. *Africanus* had clearly begun the pattern of large back molars, strongly buttressed jaws, and other adaptations for heavy chewing that reached their fullest expression in the later robust australopithecines. In light of the evidence, the simplest explanation was that *africanus* belonged at the foot of the branch leading to the later robust forms, and eventually to extinction. What we ended up with then was a simple two-pronged fork, with the Hadar-Laetoli species falling just before the branches diverged. (See illustration on page 116.)

The phylogeny complete, Tim and I had left the ticklish task of naming our common ancestor's genus and species. The genus is the more meaningful part of a new species binomen, since it links the newcomer with similarly adapted beings. For a genus designation we had three choices: *Homo, Australopithecus,* or "something else." Quickly we eliminated the last option; there weren't that many unique features about the new species, and a new genus name would completely obscure all the similarities that it shared with all the later hominids but not with the apes. For a while I pressed for *Homo,* once again on the basis of that small-toothed dental pattern. But I began to see more and more how odd that would seem, to admit into the human genus a species with a small brain capacity, a face that projected forward in an apelike manner, and a host of other primitive traits. So we were left with *Australopithecus.* Neither of us was delighted with the name, but it was the inevitable conclusion of our reasoning, the "least wrong" of three uncomfortable choices.

I knew that Mary Leakey would like it even less. She'd never cared for the term in the first place, believing that it was unwise of Dart to have named a genus on the basis of a juvenile skull. Much more to the point, by calling the common ancestor *Australopithecus,* we were declaring that the human line had to be younger than Lucy. Essentially, that would extinguish the notion of ancient *Homo* that had been passed like a torch from Arthur Keith to Mary's husband, Louis, and on to her son Richard. Unwelcome as that news would be, we hoped she would see that there was no logical alternative. For a species name, we came up with *afarensis,* acknowledging the geographical region of the Afar Triangle. We also found a way to pay respect to Mary Leakey's Laetoli site, and underscore the important connection between it and Hadar. According the International Code of Zoological Nomenclature, when naming a new species, one fossil must be chosen as a "type specimen," a sort of flagship fossil that formally represents the entire collection. As type specimen we chose L.H.-4, the fairly complete mandible from Laetoli that had already been published and that exhibited the features diagnostic of the species.

After two grinding weeks, Tim and I emerged with the bulk of our work finally behind us. We presented its outlines in a letter to Mary. She wrote back in January: Encouragingly, she conceded that it was perhaps time to name a new species, but as I had expected, she flinched at the name *Australopithecus.* By this time we knew there was no alternative, but we still felt Mary should play a part. I wrote back suggesting a plan: Rather than publish one massive article, we would first propose the new species in a *descriptive* paper, co-authored with Mary and Yves Coppens, and save the *interpretive* discussion—the presentation of our new family tree—for a separate publication authored only by Tim and myself. This would save Mary from being associated with any implication that *Australopithecus* was ancestral to *Homo.*

Mary was scheduled soon for a lecture tour in the States, and late February she stopped in at Berkeley to meet with Tim.

"You know, I don't like it," she said after they'd finished with the small talk.

"What don't you like?"

"That term, *Australopithecus.* I hate the word."

"Why do you hate it?" Tim asked.

"I don't know. I just hate it."

"I don't like it much either, Mary, but we haven't got any choice," Tim said. Then he went over again all the reasons for eliminating *Homo* and every other possible alternative. Finally he convinced her that there was no other way to go. She agreed to join us as co-author of the descriptive paper—so long as it made no mention of *Australopithecus*'s being ancestral to *Homo.* Tim and I finished writing the paper in the late spring, sent drafts off to Mary, and submitted it to *Kirtlandia,* the journal of the Cleveland Museum. We also took the precaution of sending a draft to Ernst Mayr at Harvard University, a doyen of evolutionary biology whose vast knowledge equips him to serve as a final arbiter on the naming of species. Mayr wrote back and approved the logic behind our name.

So encouraged, we wrote up the interpretive paper in May and submitted it to *Science.* Both of us knew that we had just completed what might be the greatest contribution we would ever make to our field—and one that put us on a collision course with some of the oldest, most powerfully defended ideas of human evolution. Naturally we were apprehensive, but we felt prepared to meet any criticism that could be leveled against us. In late May I left for Stockholm. I had been invited to present the Hadar fossils at a Nobel Symposium: a perfect opportunity to announce the new species and present our phylogeny to the scientific community. I had not anticipated that the occasion would also mark the end of my friendship with Mary Leakey, and the beginning of a ceaseless rivalry with her son Richard.

Both were in the audience that afternoon. I gave my lecture as planned, unfolding our reasoning step by step. It was necessary, of course, to talk about the Laetoli fossils at some length, since they played an important role in our hypothesis. I made sure that I gave Mary Leakey full credit for her discovery of those specimens and the development of the Laetoli hominid site—which by this time had also yielded up those ancient footprints. I was unaware, however, that as I talked Mary was growing red with anger. She had planned to discuss the Laetoli material herself immediately afterward and was furious that I had appropriated *her* fossils. She claimed later that she had never been informed that the L.H.-4 mandible had been designated the *afarensis* type specimen, had never consented to be a part of the naming of the species, and indeed had never even believed that there was only one species represented among the specimens from Hadar and Laetoli.

I was completely unprepared for her reaction. Mary had been sent a draft of the *Kirtlandia* paper and had never given Tim or me any indication that she had changed her mind about being a coauthor. She later held that she had never given us permission to use her name in the first place, and we were simply exploiting her reputation to bolster interest in our own theory. Richard was angry too. He had been instrumental in organizing the symposium, and he believed it was an inappropriate setting for the naming of a new species. He too accused us of exploiting the aura of the Nobel name to heighten attention to *afarensis.*

I have answered these accusations many times over the last ten years, and I needn't go into them here in detail. Richard's objections are silly; the Nobel Symposium, entitled "Current Argument on Early Man," was as appropriate a forum as any to name a new "early man." As for my discussing Mary's fossils: To cite unpublished fossils is indeed a serious breach of scientific etiquette. The Laetoli fossils, however, *had* been fully published and described two years before, by Tim. As a result, they belonged in the public domain, and I was as free as anyone else to refer to them in a talk. What really irked Mary, I think, was my citing the Laetoli specimens to support a view of human evolution that Mary did not share—first, that man was descended from an australopithecine, and second, that humanity was not nearly so old as the Leakey tradition maintained. The split-publication scheme that Tim and I had developed with her consent was intended to keep her dissociated from those ideas. Apparently it wasn't enough.

To add another twist, Tim was due to spend the field season that summer with Mary at Laetoli.

"She was running an anti-Johanson campaign in camp that summer," Tim told me afterward. "She kept asking me, 'How can you associate with *that man*?' I told her that if she felt so strongly about it, she ought to get her name off the *Kirtlandia* article before it was published."

Mary sent a telegram to that effect to my office in Cleveland. Though the journal issue had already been printed, we were able to halt shipment and reprint the title page of our article without Mary's name on it. A few months later, the interpretive paper presenting the Johanson-White phylogeny appeared as a cover article in *Science*. *Australopithecus afarensis* was out in the world for all to behold. A front-page story quickly followed in *The New York Times*. Suddenly I found myself swamped by requests for interviews and television appearances. It was an exhilarating time. But amid all the hoopla I remained acutely aware that in Mary Leakey, the birth of *afarensis* had cost me a treasured and respected friend. I worried too about how long my cordial relationship with Richard could last.

On the rim of the great Ngorongoro Crater there is a wildlife game lodge with a magnificent view of the crater floor two thousand feet below. The Olduvai camp was only an hour's drive down the other side of the extinct volcano, but we decided to stop at the lodge for a quick cup of tea before heading out on the last leg of the journey. The lodge possessed a refrigerator—the nearest place to the dig

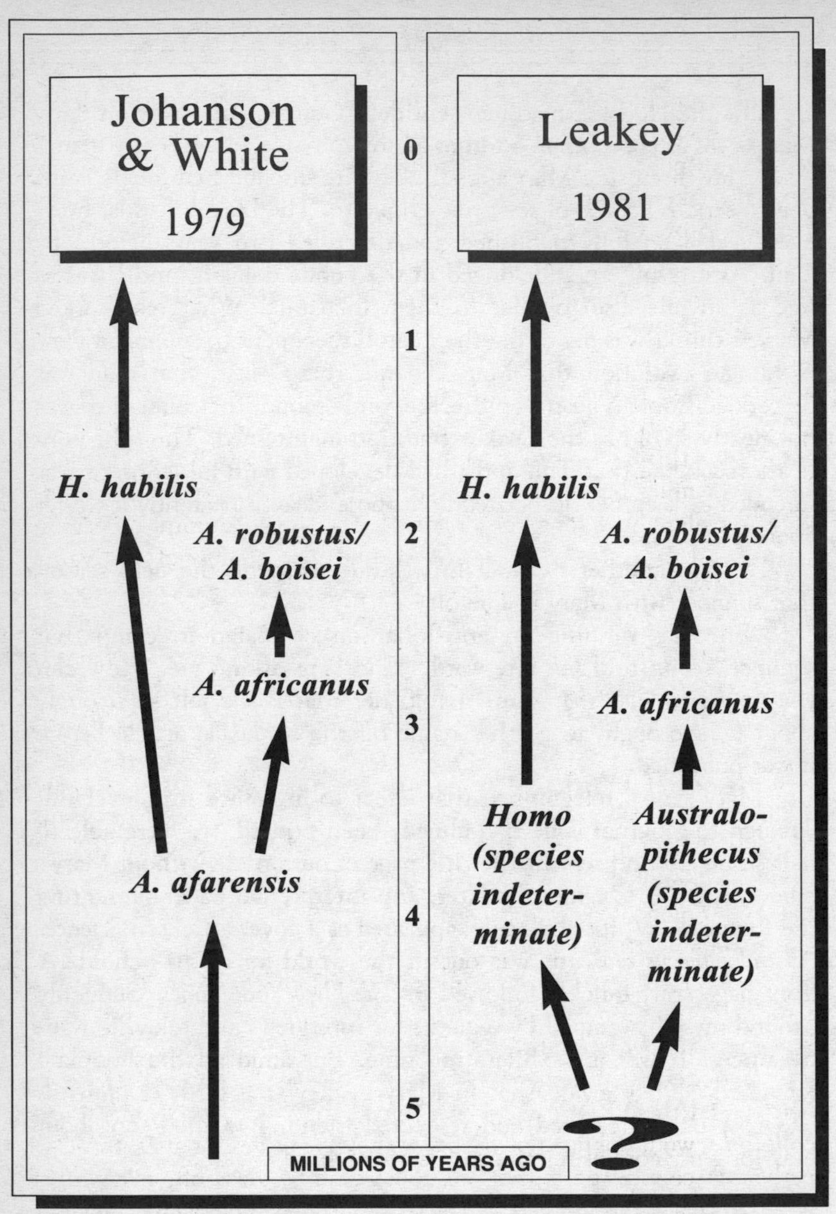

The human family tree proposed by Johanson and White in 1979, compared with another interpretation given by Richard Leakey in 1981. Johanson and White believed A. afarensis *was the ancestor of all later hominids, including* Homo. *Leakey contended that they mistakenly yoked two separate species under the name* afarensis, *and that the true ancestor of man had yet to be found. Compare these trees with those on page 132.* DESIGN BY VAN HOWELL AND DOUGLAS BECKNER.

where we could keep fresh the snake antivenin we'd brought along. There was little threat of poisonous snakes at Olduvai, but Laetoli was full of puff adders hidden in the grass. The fangs of these thick-bodied snakes deliver a highly toxic venom rendered all the more pernicious by the speed by which it takes effect. Untreated, a bitten arm or leg can turn black and swell up to twice its normal size within minutes; in a few hours internal bleeding sets in, and there is little left to be done to save the victim. I started calculating how long it would take us to rush a bitten team member back up to the lodge from Laetoli on the other side of the Gorge. Then I decided there were some calculations better left unfinished.

From the lodge the road hugged the rim of the crater for another few miles before beginning its descent to the Serengeti Plain. The prevalent mist on Ngorongoro keeps the mountain capped in green, and on either side of the narrow road patches of cultivated crops were wedged in gullies and ravines. Children waved and shouted to us. An occasional Maasai warrior, face and hair coated heavily in ocher, stared impassively at the black and beige van as we passed. Soon the road dropped down onto the shoulder of Lemagrut—another of the extinct volcanoes whose ancient eruptions laid down some of Olduvai's famous beds. All trace of green disappeared. As we settled over a ridge the Gorge appeared teasingly in the distance, a slight, darkened fold pressed into the edge of the dust-dry plain. We'd been chattering through most of the trip, but now all four of us were silent. Now so close to the place that personified the Leakey spirit, I could not help but feel nostalgic for the times when I came this way to visit Mary for a few days of restful conversation, or spent afternoons sailing in the Indian Ocean in Richard's boat. How odd, I thought, that now I was welcome at Olduvai only because Mary was gone, taking her furniture with her.

In spite of the personal irritation that the naming of *afarensis* had caused them, Tim and I had hoped that the Leakeys and their colleagues would oblige our theory with some professional response—perhaps a counterthesis that we could bounce ideas off, even integrate into a more refined synthesis. Our new phylogeny, like all family trees, was not some rigid truth, but rather an interpretive tool—a hypothetical structure meant to help organize thinking. A theory like that only gathers strength as it is tested and retested by competing ideas. I was particularly anxious to hear what Richard would have to say. A couple of weeks after our announcement in

January 1979, he arrived in the United States for a lecture tour. I met him at a symposium in Pittsburgh.

"I think Don was right about the Hadar fossils the first time," Richard said at the symposium, a reference to my 1976 paper with Taieb. He believed that there were at least two species at Hadar, one of them "some kind of primitive *Homo.*" His arguments, though, were exactly the ones I had used initially against Tim, so I was prepared to counter them. I even offered to review the fossils with him, point by point. Richard declined.

Finally, a year later, both Mary and Richard responded to *afarensis* in print. Tim and I were amazed: Once again, we had hoped for something substantial we could sink our teeth into. Instead, Mary Leakey and two colleagues attacked us in *Science* only on formal, nomenclatural grounds, offering a thin handful of objections to our choice of name, type specimen, and so forth. Many of their objections centered around the fact that a reporter for a British popular magazine had attended the Nobel Symposium and filed a story on *afarensis* before the name was formally announced in a professional journal. Technically, the magazine story was thus the formal announcement of the species, and since the reporter had ignored various rules of nomenclature, the species name was suspect! Richard Leakey's accompanying response, co-authored with Alan Walker, was almost as meager, challenging the validity of *afarensis* on the basis of a confusing weave of phylogenetic arguments that failed to address the morphology of the specimens at all.

For three years, that was all we heard from the Leakeys. In his public talks, Richard would give only a sprinkling of statements about *afarensis,* always unsupported, implying that Tim and I were mistaken, and that more fossils were needed before anything could be further clarified. I felt I was fencing with a phantom.

"I'm so sick of hearing about the need for new fossils," Tim said to me once in the summer of 1981, soon after I'd moved to Berkeley to establish the Institute for Human Origins. By this time he was a professor at Cal. "If new fossils prove us wrong, *fine.* I hope Richard finds them. But for once, lets see him put his cards on the table."

Tim and I were sitting in a café just across from the campus. *Lucy,* my book written with Maitland Edey, had recently been published, and I had been invited to appear on Walter Cronkite's *Universe* television series. The producers told me that Richard had also agreed to be on the show, and asked if I had any objections. Far from objecting, I was delighted. Richard had been claiming that the "rivalry" between us was a myth, largely the invention of the press. I thought that was misleading, and I welcomed the opportunity to meet with Richard on the record.

"This could be your only chance," Tim was saying. "You've got to press the point that there are major differences in the way we draw the family tree."

The broadcast was scheduled to be taped at the American Museum of Natural History in New York. The producers had asked me to bring along some props, so I carried with me a gorilla skull and a cast of an *afarensis* skull that had been carefully reconstructed by Tim and Bill Kimbel from pieces of several individuals. I had also put together a chart: On one side was the Johanson-White family tree. I'd left the other side blank.

I got to the museum with about half an hour to spare. Richard arrived about fifteen minutes later and greeted me in a friendly fashion. He told me that he too was about to publish a new book, the companion volume to the television series *The Making of Mankind.* He had with him some foundation people, his publisher, and his literary agent. Several paleontologists from the museum were also gathered around. Richard guided me behind a partition.

"Don, I hope this doesn't turn into a debate," he confided. "Our differences are really of a minor nature, and it would just confuse the public if we started going on about jaws. There's a terrible threat of creationism in this country, and we should show a united front."

"I don't agree that our disagreements are minor, but I don't know what's going to happen." I said. "I suppose it's up to Cronkite."

A few minutes later Cronkite appeared and we introduced ourselves. He was something of a childhood hero for me, and I was thrilled to meet him. Cronkite explained how the interview would unfold—first a question to me, then one to Richard, and so forth. Richard told him the same thing he'd said to me—that he didn't want the program to turn into a debate.

"I'm not Dan Rather and this isn't *60 Minutes,*" said Cronkite.

The three of us sat down in a makeshift set constructed in the paleoanthropology section of the museum. In the background was a display of skulls. I put my family tree chart beside my chair, and the *afarensis* skull on the table in front of me. As soon as the cameras started to roll, it became apparent that a debate was just what Cronkite was looking for.

"We brought Donald Johanson and Richard Leakey together in the American Museum of Natural History to discuss their different ideas of man's ancestry," Cronkite said. He gave some background information on 1470, Lucy, and the Johanson-White hypothesis. Then it was my turn to speak. I laid out the situation as clearly as I could.

"There has been a controversy that has been going on now for nearly three years between Richard and myself, and it specifically focuses on the family tree," I said. I explained that we had presented our phylogeny in January 1979, but that others, specifically Richard, believed that it did not fit the evidence. Leakey's turn.

"I've heard it all before," he said. "I think it is marvelous. I just don't agree . . . but I'm not willing to discuss specifics. . . . [I'd heard that before too, I thought.] I'm not going to say whether you are right or wrong. But I think you're wrong." Then he gave a little laugh, as if the whole thing was in good fun and we should now move on to other questions. But I wasn't going to let him off so easily. My turn.

"I've brought along a sort of a portrayal of how I look at the family tree," I said. And with that I reached back and took out my chart. "And I've left a spot for you to draw your rendition."

"No, I don't think we can do this," Richard said obviously flustered. "I haven't got crayons, I haven't got cutouts, I'm not an artist . . ."

"I thought of that," I said, and I took a felt-tipped marker from my pocket and put it on the table in front of him. Richard left it lying there and said nothing. But as I started to explain the crux of our differences, he suddenly picked up the marker and asked me to hold one edge of the chart.

"I think in probability I would do *that,*" he said, scrawling an angry "X" over my family tree. Now it was my turn to be taken aback. Meanwhile Cronkite was practically squirming with delight.

"And what would you draw in its place?" I said.

"A question mark!" Richard replied, and he drew one in the empty space. Then he sat back in his seat and laughed.

My turn once more. I explained how Richard's family had always believed that man's roots went back millions and millions of years, and that *Australopithecus* had nothing to do with our ancestry. Then I elaborated a bit on finding Lucy, and how we believed her to have everything to do with our origins. Richard had more chances to respond, but each time he dodged. Instead, he talked about how much he'd like to see new fossils discovered.

"I would love to prove him right," he said, "but I might just prove him wrong." One more little laugh, and the debate was over.

As soon as the taping was finished Richard got up and hurriedly left the set. I was told afterward by one of the producers that he refused to sign a release on his way out, declaring that he would not allow the show to be broadcast. Leakey's lawyers called and threatened to sue if the tape went on the air. According to the producer, Cronkite himself had to intervene, apologizing for any misunderstanding but making it clear that the broadcast was news and did not require any release. The show was broadcast as planned. Later, Richard claimed that he did not know about the shoot until the very day of the taping, and that Cronkite and I had planned the whole thing together. Neither statement is correct.

When I next saw Richard, he looked right past me. That was four years ago. We haven't spoken since. In public, he has continued to insist that our "rivalry" is largely the media's creation, and that beneath the hype there lie nothing more than minor professional disagreements. I wish it were that painless.

Through the 1980s, Richard expanded his search for ancient Man to the other side of Lake Turkana from Koobi Fora. The West Turkana region has yielded a lode of new finds—most recently the surprising "Black Skull." Richard should be commended for his part in that discovery. But neither the Black Skull nor any other fossil found in the last ten years lends an ounce more support to his belief in the antiquity of the human line. Richard has stuck to the family tradition nonetheless, believing that further exploration will eventually prove his father right.

It's his privilege, of course, to believe anything he wants, and someday he may well find the proof he seeks. For my part, I am quite happy not having a legacy to uphold. My own father—a barber by profession—died when I was two; I scarcely even remember him. I'm sure I bring prejudices of my own to the worktable when I examine a fossil, but I'd like to think that my belief in a more recent birth to humanity has arisen from the nature of the evidence that we have in hand. Nothing—not even the loss of friendships once held dear—can weaken that conviction. Nothing, that is, except for new evidence. And I will search for it whenever and wherever I am given the chance.

Not all of the critics of *afarensis* turned out to be as evasive as Richard Leakey. Though the 1980s, the legitimacy of the new taxon, and our conviction that there was only one species represented at

Hadar and Laetoli, was challenged over and over again, by seasoned investigators whose own views were threatened by it, and by young Ph.D.s cutting their teeth on a vulnerable target. Phillip Tobias, who had assumed Raymond Dart's chair at the University of Witwatersrand, was one of the first to respond. If fossil species can be "insulted," it was Dart's *africanus* who had suffered the worst affront in our reorganization of the human line: In effect, we had booted it out of its seminal role as the common ancestor of all known hominids, and shunted it onto the robust australopithecine line headed toward extinction. Tobias picked away at our published descriptions, trying to show that *afarensis* was in fact no more than an East African variant of *africanus*. In 1983, we responded with a detailed review of the facial, dental, and skull traits that distinguished *afarensis* as a separate, more primitive species. Tobias was eventually convinced enough to include the species in his own phylogenetic reconstructions.

Most of the other critics of *afarensis* bridled at the notion of one speicies yoking all the Hadar and Laetoli specimens. In each case, the scientist would focus in on a collection of anatomical details and point out how very different one group of Hadar fossils looked from another—at least in the specifics of anatomy that scientist had chosen to focus upon. Yves Coppens, who had joined Tim and me in the original announcement of *afarensis*, changed his mind and declared that he believed there were two species at Hadar after all, citing some differences in the cusps of premolars. His students Christine Tardieu and Brigitte Senut homed in on some differences in the shape of the femurs and upper arm bones of the large and small Hadar specimens. Adrienne Zihlman of the University of California at Santa Cruz declared that the difference in size among the arm and leg bones was alone too much to bear grouping into one species. Meanwhile, Dean Falk of Purdue University scrutinized the Hadar skull fragments and announced that the patterns of blood-drainage channels, faintly perceptible on the inside of the fragments, fell neatly into two groups—one looking like *Homo*, the other resembling those of the robust australopithecines.

Our most vociferous detractor was Todd Olson, a young anatomist at the City University of New York. Olson built his case substantially on a difference he perceived among the Hadar skulls in a region called the mastoid—the hump of bone that you can feel projecting just behind your ear. According to Olson, *some* of the Hadar mastoids were puffed out—like those of the robust australopithecines—while others resembled the flatter condition of *africanus* and *Homo.* Based on his interpretation of the mastoids and a selection of other features, Olson was convinced that Lucy and some similarly gracile fossils from Hadar represented the first members of the human line, a species he called *Homo aethiopicus.* The other Hadar hominids belonged on the robust lineage.

Tim, Bill, and I responded to each of these critics in turn, even when we felt that their objections were trivial or had been addressed before. The bottom line of our counterargument was the same one we would draw for anyone discriminating species differences on the basis of a few carefully chosen traits: No two animals have the exact same anatomy, even if they are members of the same population, let alone the same species. Meanwhile, through all the salvos launched against *afarensis* in scientific meetings and journal pages, no one had ever aimed a shot at the whole corpus of facial, dental, and jaw characteristics that had brought home to us the clarifying unity of the Hadar and Laetoli fossils in the first place. If people ignore the evidence you marshal to support an argument, and instead strain to find *other* evidence proving you wrong, then you start to think you've made a pretty good case.

Collectively, the anti-*afarensis* attack was further weakened by a lack of coherence: Olson and the others all were shouting "two species at Hadar!," but they did not agree on which fossils belonged in which of the two groups.

Nevertheless, by that summer of 1986 the mere fact that *afarensis* had drawn such persistent attack seemed to keep the skeptics going on sheer inertia. Or perhaps something more powerful than inertia was fueling all the quibbling over femur sizes and blood-drainage channels. Perhaps it was the old hope for the antiquity of *Homo,* threatened when Tim and I drew the point of divergence between the two branches of the hominid family tree *after* Lucy's time. If we were wrong and there really *were* two species at Hadar in the late Pliocene, then their lineages could again be extended back in time before three million years, to meet at some distant point in a still-to-be discovered common ancestor. The unconscious unspoken subtext of that hope is a belief in the venerable uniqueness of the human line, distinct from the rest of creation, even as it fades from sight into the unwitnessed past. Such a powerful idea is not going to die just because a couple of scientists read some bones and come up with a different story. After all, it is what humanity has believed about itself for centuries.

The year before, Richard Leakey's team at West Turkana had come up with a new fossil that could be construed to keep that hope alive. "The Black Skull," as it came to be called, was found by Richard's long-time colleague Alan Walker. Alan had brought a cast of the skull with him to a scientific meeting just a few months before we had left for Olduvai Gorge. I had not been able to attend the meeting, but I asked a colleague about it later.

"Walker was the happiest man in town," my colleague told me. "He was carrying that skull around in a blue plastic bag, which he would open up for anyone who wanted a peek. That bag drew people like candy draws kids."

Alan Walker had every right to be proud of his find. For years paleoanthropologists, myself included, had been lamenting the lack of fossil evidence for human evolution from the critical time period two to three million years ago, between the known presence of *Australopithecus afarensis* and the appearance of all later hominids, including *Homo.* Alan's discovery was securely dated smack dab in the middle of that chronological "black hole." It was a beautiful specimen, with much of the braincase and face intact. Only one complete tooth remained; it had fallen out of the upper jaw. The biggest surprise about the Black Skull, or KNM-WT 17000 as it is formally named, was its baffling mixture of very primitive and very advanced traits.

In taxonomy, the word "advanced" does not mean "better," and it certainly does not mean "more human." Advanced traits are simply those that are more specialized—evolved away from a primitive, generalized condition. A less misleading synonym is "derived." Even as Walker was gathering the pieces of the Black Skull off the ground, he could tell that this new specimen was derived in the direction of the robust australopithecines like Zinj, especially in the sheer massiveness of its face and jaws. The one preserved tooth, a premolar, was also enormous, similar to Zinj's own.

But in some other respects—especially in its hind portions—the skull seemed remarkably primitive, bearing the strongest similarities to *A. afarensis,* the most *unspecialized* hominid known. The base of the cranium, for instance, was flat rather than flexed, a primitive condition related to the apelike forward jut of the fossil's face. The size of the braincase indicated a minuscule brain capacity—only 410 cc, smaller than any adult hominid known save one partial specimen from Hadar. To accompany its teeny brain, the skull sported a bony ridge running down the back.

Clearly, this fossil was no ancient *Homo,* and on its own could add nothing to the debate over the number of species at Hadar. But many of the scientists who scorned *afarensis* nevertheless greeted the Black Skull as if it was some samurai warrior come to do battle for their cause. They argued that if the new cranium looked like a derived robust species in the face, but a primitive *afarensis* in the back, maybe that was because *afarensis* was merely a chimera.

At issue was the legitimacy of the *afarensis* skull reconstructed by Tim and Bill—the same one that I had brought to the taping of the Cronkite show several years before. We had never found an intact skull of a single adult at Hadar. To fill that important gap, Tim and Bill had painstakingly pieced together some face and cranial fragments from several Hadar individuals of approximately the same size. No one questioned that the reconstruction had been masterfully handled, but the fact that it was an acknowledged composite left it vulnerable to attack from the two-species adherents, including Richard Leakey. True to form, Richard did not offer his views in print. But in public lectures given soon after the announcement of the Black Skull in 1986, he insinuated that the reconstructed skull was nothing more than a hodgepodge conglomeration of two species.

"I know that Don would not like to agree with my interpretation," he told one audience at Cal Tech, "but science is not the business of agreeing or not agreeing."

Alan Walker's wife, Pat Shipman, also a paleoanthropologist, made the case explicit in the pages of *Discover* magazine. She proposed that Tim and Bill might have force-fit the braincase of an early robust australopithecine on to the faces of more gracile forms. This mistake would explain why the robustlike Black Skull matched up with *afarensis* only in the back, not in the front. Presumably, if an expedition was allowed back into Ethiopia and was lucky enough to find a bunch of new skulls, some of them would look robust both back *and* front, while others would be gracile all around. Two species, not one. Out of the ashes of *afarensis,* the hope for ancient *Homo* would emerge once again.

"Here we go again," I said to Tim when he showed me the *Discover* article. "Tell me, what sort of strange geological circumstances do you think would have destroyed only the faces of the

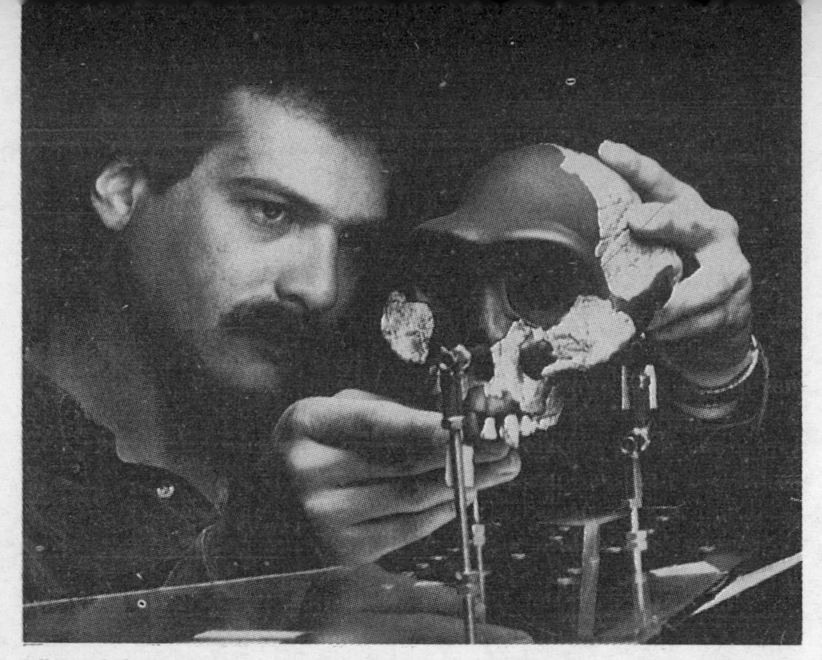

Bill Kimbel studies the skull of afarensis *he reconstructed with Tim White. No complete crania or skulls were found at Hadar, but Kimbel and White's composite, based on fragments from twelve male individuals, gave researchers a tool for better understanding the anatomy of* afarensis. DAVID BRILL

alleged robust species, and only the braincases of the gracile one?"

There were other problems with Shipman's argument. She claimed that there was no specimen in the Hadar collection of skull fragments in which continuous bone connected the face to the braincase—a bit of evidence that would help determine the accuracy of our reconstructed composite skull. She was simply mistaken on that point. One Hadar specimen, A.L. 58-22, is indeed comprised of both facial and cranial pieces. There is nothing about that specimen that shows it is derived in the direction of the robust australopithecines; on the contrary, it is a perfect example of the primitive, generalized morphology that led us in the first place to believe that *afarensis* was the ancestor to all later hominids. Most important, among the hundreds of fossils in the Hadar collection, we had never located a single anatomical trait that could unequivocably place a specimen on the robust australopithecine lineage.

The Black Skull notwithstanding, today almost every one of our critics has come around to accepting *afarensis*. There is no question, however, that Alan Walker's extraordinary find demands a redrawing of most versions of the family tree—including the one Tim and I had

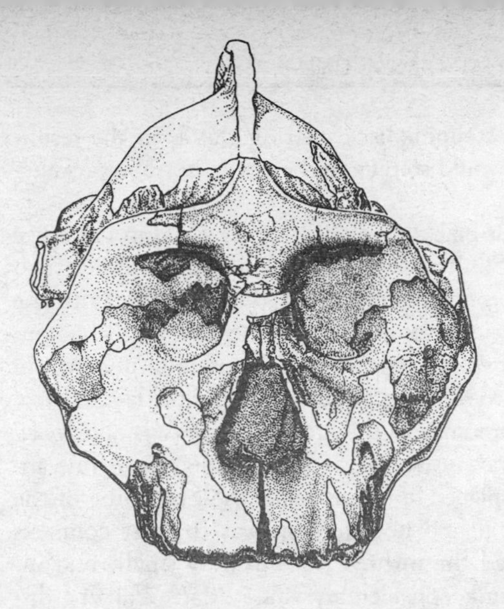

The enigmatic Black Skull (KNM-WT 17000), found in Kenya by Alan Walker in 1985.

proposed back in 1979. In our original accounting of the known hominids, we had seen the very specialized, robust characteristics of East African *boisei* and South African *robustus* as the end products of a trend in australopithecine evolution. This trend toward massive back teeth and big jaw muscle attachments could be seen faintly in the primitive *afarensis,* became progressively amplified in *africanus,* and reached its fullest expression in *robustus* and especially the "hyper-robust" *boisei.*

In our analysis, we had placed *boisei* in the same species as *robustus,* although evidence was mounting even then for keeping them separate. Yoel Rak, a paleoanthropologist at Tel Aviv University, made a particularly strong case for a species-level distinction for *boisei* in 1983, based on a detailed analysis of australopithecine faces. The Black Skull further strengthened this distinction, defining a separated evolutionary history for *boisei* in East Africa. Indeed, in their analysis Walker and Leakey asserted that the 2.5 million-year-old specimen could be assigned to the species A. *boisei.*

We didn't think things were quite that simple. In the basement lab of the Institute of Human Origins, there is a blackboard set along the length of one wall. Before leaving for Olduvai, Bill Kimbel and I had started sneaking down to that blackboard whenever we could find a spare moment. Sometimes Tim would take a break from his duties at the university and wander over. Anybody looking down on us through the sidewalk-level window would have seen the same scenario repeated again and again: a man stepping up to the blackboard, scribbling something, and stepping back, staring glumly at the result. Then he or another man would step up and erase it, scribbling something else in its place.

On the left side of our blackboard, Bill had chalked in the three most popular pre–Black Skull versions of the hominid family tree. All were essentially two-pronged family trees, differing chiefly in the placement of *Australopithecus africanus*. In the one Tim and I had put forth in 1979, *afarensis* was depicted as the last common ancestor of all subsequent hominids. One branch of our tree led directly to *Homo*, while the other led from *africanus* to the "robusts," *robustus* and *boisei*. A second version had *africanus* shunted over to the *Homo* line instead. A third phylogeny placed *africanus below* the branching of the forks of the tree, so that it and not *afarensis* was the last common ancestor. Bill had sketched these three trees in very small, and indeed their value had shrunk considerably since Alan Walker's discovery had burst upon the scene.

Next, we made a list of the various distinguishing traits of the Black Skull. The rest of the blackboard was reserved for experimenting, trying to squeeze some sense out of the baffling problems the new fossil presented. At this time we were also beginning to formulate ideas for a paper we were to present at a symposium on the evolutionary history of robust australopithecines, organized by Fred Grine at the State University of New York at Stony Brook. We decided to take the opportunity to see just where the Black Skull might fit in the scheme of things.

Like all taxonomists attempting to establish the relationships between extinct species, we were forced to work around a critical "as if": the assumption that a morphological similarity between any two fossils implies an evolutionary relationship. At the same time, we knew that such similarities can often be deceiving. To take just one obvious example, sharks and porpoises both have fins, torpedo-shaped bodies, and numerous other adaptations for swimming, but no one would suggest they are closely related. Porpoises are mammals, of course. The fact that they share some very specific adaptations with sharks is an example of *evolutionary convergence*—two unrelated species converging on similar solutions to an environmental problem.

The very different evolutionary histories of sharks and porpoises are easy to see when you look at them as whole organisms, with many complexes of traits that have little to do with each other. But what if you only had two fossilized dorsal fins to go by, from animals

Paleoanthropologists Alan Walker, Yoel Rak, and Bill Kimbel compare a cast of the Black Skull, left, with that of another robust australopithecine. DONALD JOHANSON

long extinct? It might not be so easy to tell whether or not they represent related species or not. The same goes for extinct hominids. And convergence is not the only pitfall waiting to confound the taxonomist. Sometimes, the same traits can evolve independently from the same starting point—what we call *parallelism.* Old World and New World monkeys, for example, look and behave in surprisingly similar ways. But their evolutionary histories have been kept separated by an ocean for at least forty million years; the traits they share have evolved "in parallel." Another curve ball that evolution can throw is called *reversal.* In this case, a trait that has been lost reappears later on in the same lineage. Perhaps a long-tailed monkey species evolved into a short-tailed one, only to sport a long tail again when new selective pressures demanded it thousands of years later.

The taxonomic monkey wrenches of convergence, parallelism, and reversal have traditionally been treated by paleoanthropologists with a kind of half-embarrassed sleight of hand: We trot them out when we need something to explain an anomaly in an otherwise neat and tidy hypothesis, but otherwise we would rather pretend they don't exist. In a sense, we have to proceed as if they do not exist, though we know that of course they do. This deliberate self-deception is excused by the scientific principle of parsimony: When considering competing solutions to a problem, the one closes to the truth will probably be the simplest. The "best" version of human phylogeny would thus be the one with the shortest list of special pleadings needed to account for the known evidence.

More than anything else that the Black Skull told us, it brought home the realization that there are *no* parsimonious versions of human phylogeny. We began our evaluation of the new specimen by comparing some thirty-two anatomical traits found in the Black Skull and other species of *Australopithecus.* Immediately, we could see that there were problems with calling the skull *boisei,* as Walker and Leakey were maintaining. Twelve of the traits seen in the Black Skull were shared with both *robustus* and *boisei,* establishing the "robust" nature of the skull, but failing to define an exclusive affinity with either one. Indeed, out of the thirty-two traits, only *two* were exclusively shared between the Black Skull and *boisei.*

At the same time, twelve of the remaining traits were primitive ones tying the Black Skull to *afarensis.* So, with its large teeth, sagittal crest, and other "robust" characters, the Black Skull was evolving in the direction of *boisei*—but it still possessed far too many primitive features typical of *afarensis* to be assigned to *boisei.* Indeed, it looked like a good candidate for an evolutionary intermediate—a "missing link," if you will—between *afarensis* and the robusts. Its 2.5-million-year-old date also placed it at the chronological midpoint between them, at 3.0 and 2.0 respectively.

The next consideration was what to call the Black Skull. One possibility was *afarensis,* but in spite of the many resemblances, assigning the new speciment to *afarensis* would obscure its evolutionary meaning. For a name, we turned to a nearly forgotten mandible from the Omo, found in 1967, which had been interpreted by Camille Arambourg and Yves Coppens as being more primitive than *A. boisei.* They called it *aethiopicus.* Like the Black Skull, the Omo mandible was dated to about 2.5 million years. We believed that the Black Skull belonged to the same species, and thus should be called *Australopithecus aethiopicus.*

Now came the fun part—redrawing the family tree. First of all, we realized that *aethiopicus* could qualify as an ancestor either to *robustus,* to *boisei,* or to both. Second, while we had previously believed *africanus* to be the last common ancestor to the robusts, it was obvious in light of the Black Skull that this was no longer a viable arrangement. The Black Skull shared too many primitive traits with *afarensis* to *africanus* to *aethiopicus* would impose too heavy a load of evolutionary reversals—the loss of a host of primitive traits in *africanus* that would make a perplexing reappearance in *aethiopicus.* The Black Skull *had* to be placed between *afarensis* and the robusts—

which unfortunately left *africanus* floating in limbo.

After lots of head scratching and chalk dust, we came up with four alternative family trees, illustrated on page 132, which we presented at the Stony Brook meeting. Not to go into excruciating detail, but all of these phylogenies imply the evolution of certain traits in parallel. Of the four phylogenies, I favor the first. In this version, the old two-pronged Johanson–White tree has sprouted a third branch directly from its fork. *Afarensis* is still the common ancestor of all other hominids, but its descendants evolve in three directions: one toward *Homo;* one through *aethiopicus* to *boisei;* and one through *africanus* to *robustus.* Tim prefers this phylogeny as well. Bill, on the other hand, likes the second tree better, with *africanus* the direct ancestor of *Homo* and *aethiopicus* the common ancestor of both *robustus* and *boisei.* The other two trees are probably less likely candidates, especially the one where *africanus* leaves no descendants.

At the time I am writing this, however, nobody really places a great deal of faith in *any* human family tree. In that sense, it may seem that the Black Skull has led us from clarity into confusion, from conviction into doubt. Such thinking is profoundly misleading. The Black Skull does what the best kind of evidence is supposed to do: It gives unambiguous proof or refutation to existing theories, and in so doing takes us one step closer to the absolute truth. To me, the realization that the truth is more complicated than we thought only makes the search for it more thrilling. Before the Black Skull, it was easier to believe that the ancestral human line was somehow distinct in its humanness, treading an inevitable and inviolate route through time unlike the paths followed by even our closest relatives. That is no longer the case. The very traits that we have assumed to be uniquely human may in fact have evolved in quite separate lineages, giving rise to ancient cousins much more like our own ancestors than we have always believed.

Some of my colleagues—notably Richard Leakey and Alan Walker—contend that until more fossils are found, we should not be wasting our time proposing family trees in the first place. I do not agree. As scientists, our first responsibility is to squeeze the last drop of information out of every bit of evidence we have in hand. But we must also be willing to take the next step, and build from that information theories that will be ready when the next discovery comes along to test their strength. That's what doing science means. Frustrating as it is, the distantly tantalizing truths about our origins will

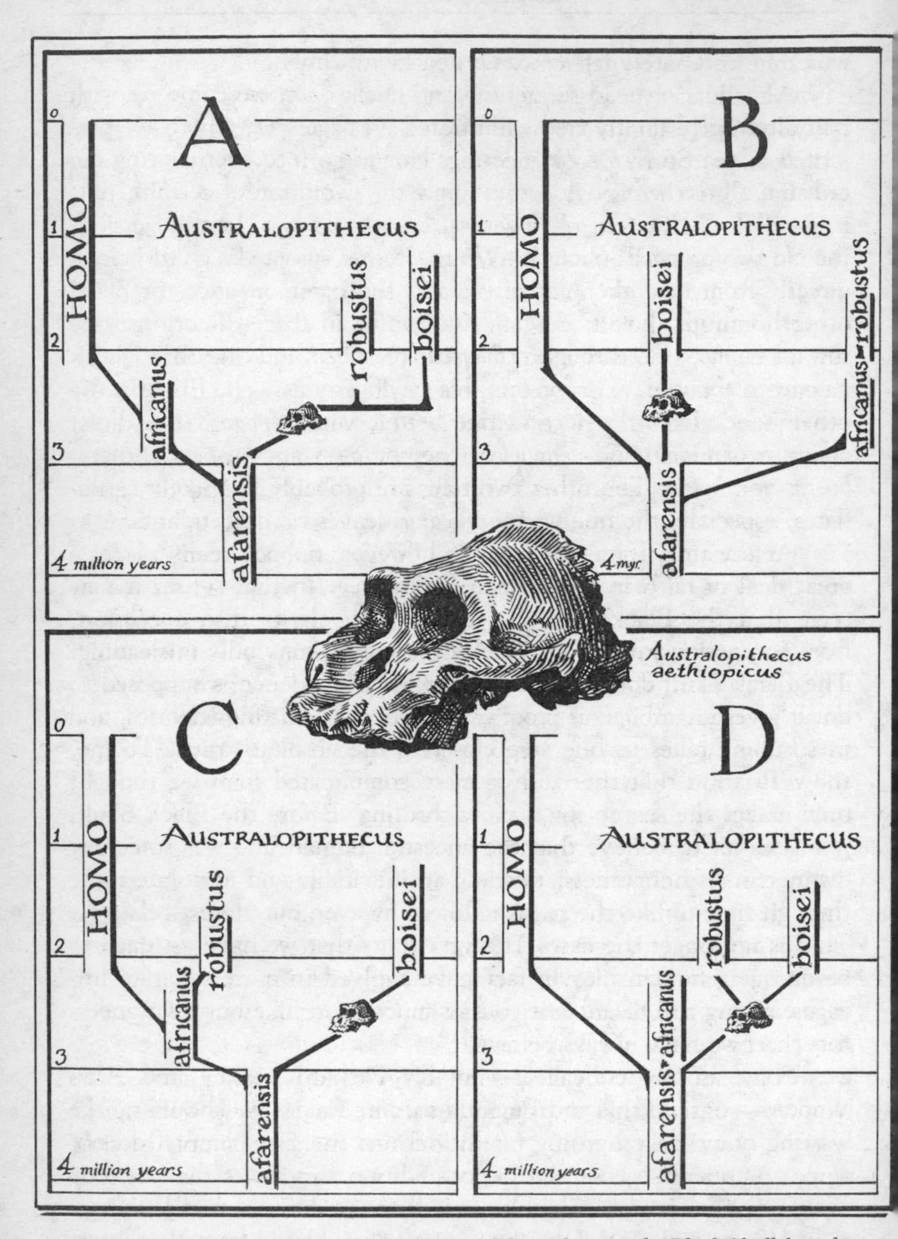

With its curious combination of primitive and derived traits, the Black Skull forced a rethinking of all proposed human phylogenies. Four possible family trees are compared here. Version B above is probably the most popular, but most scientists agree that the evidence now available does not clearly support any one phylogeny over the others.

probably not be revealed before we ourselves are buried under the earth. But that will not stop me from testing and retesting new hypotheses, exploring further possibilities. The point is not to be right. The point is to make progress. And you cannot make progress if you are afraid to be wrong.

In the summer of 1986, my immediate concern was to find more fossils. We arrived at the Olduvai camp close to 4:00 P.M., the A&K combi surprising a trio of giraffes from a stand of acacia just outside the gate. They floated away with the heart-stopping grace giraffes alone possess, and we drove on through. Some of the camp staff were out in the yard unpacking cooking equipment. There was smoke coming from the cook's shed, and at least one of the two windmills was spinning steadily in the breeze above the Gorge. Gerry Eck came out to welcome us with relatively good news—though Mary had taken the iron stove, Stanley, the cook, had rigged up a serviceable wood-burning stove from a metal drum. Meanwhile Gerry had tended to the one working windmill and had it producing a modest current. He told us that the rest of the expedition members were out surveying in the main Gorge near FLK. Gen and Berhane are by temperament a lot less excitable than I am, but now they were fidgeting like two kids who'd just been dropped in the middle of Disneyland. I knew Tim was exhausted from jet lag, but he agreed anyway to take them down to meet up with the others.

I didn't want to go with them right away. After dropping my bag in what had been Mary Leakey's hut, I walked out across the yard, hopped over the fence, and kept walking in the direction the giraffes had floated, out toward the dusty level plain that stretched beyond to Seronera. Now that we'd finally reached camp, a feeling of uneasiness had taken possession again. What were we really doing here? What could we hope to accomplish in six short weeks? For whatever reason, there were people who I knew would be more than happy if we accomplished nothing at all. Maybe they were right. Maybe Olduvai was a poor proposition, and it should simply be shut down and declared a Tanzanian national monument, as Mary herself supposedly believed. On what presumption did I think the Gorge still held any promise?

Maybe the giraffes had the answers. A hundred yards off to my left I could see them browsing above the thorn trees—"long-stemmed, speckled gigantic flowers," Isak Dinesen called them—their tapered necks kneading the air as they floated tree to tree. They seemed so completely remote from the very notion of worry that I suddenly felt a pang of envy, as if it were somehow a better fate to face the lions of the Serengeti than to struggle along in the anxious environments we create with our own predacious ambitions. I turned around and looked back at the camp buildings, surrounded all around by a fence heaped with thorn branches to keep out lions and other intruders. Olduvai Gorge was my workplace now, though not under circumstances I could have foreseen. We would have to wait and see where that would take us.

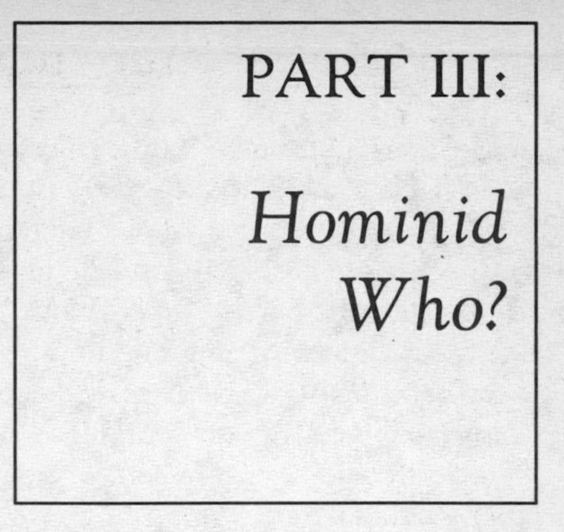

PART III:

Hominid Who?

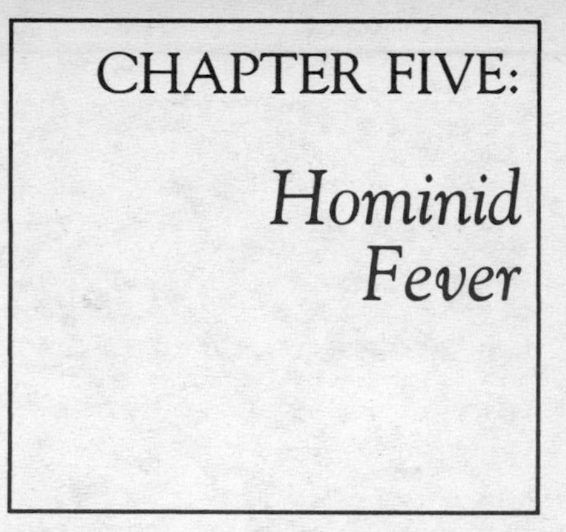

CHAPTER FIVE: *Hominid Fever*

But who knows the fate of his bones, or how often he is to be buried?
—Sir Thomas Browne

A THORN TREE STANDS IN front of the central building of the Olduvai camp, its umbrellalike canopy nearly touching the ground. Years ago Mary Leakey had a stone birdbath built beneath the tree. In the middle of the afternoon of July 21, I was sitting at the table in the open-air common room, watching a flock of weaver birds take turns at the bath. The acacia seemed ripe with chattering, bright yellow fruit. Every few moments a bird would slip off its perch and join the crowd already splashing around in the bath below, squeezing another bird back up into the tree. The weavers were all that moved in my range of sight; everything else was wrapped in the hazy stillness of midday. By the camp road—two parallel tracks of matted grass—a trio of goats dozed in the shade of a smaller acacia. The road passed through the camp gate, buried itself in a thicket of whispering thorn, and descended unseen into the Gorge. On the far side, the Serengeti swelled up toward the ancient eroded slopes of the volcano Lemagrut. It was the season when the local Maasai herdsmen burn the grasslands to encourage new growth, and a tulle of suspended smoke and dust veiled the volcano's summit.

We had been in the Gorge now for three days—time spent familiarizing ourselves with the site, with the work that had to be done,

Members of the 1986 Olduvai Research Expedition. Top: Paul Manega, Bob Walter. Middle: Tim White, Donald Johanson, Prosper Ndessokia. Bottom: Berhane Asfaw, Gen Suwa DONALD JOHANSON

and with each other. There were fourteen of us in camp, including six scientists (two accompanied by their wives), four graduate students, and a couple of visitors. Field expeditions make for strange bedfellows; out in the middle of nowhere you throw some people together who may have little in common, and you hope that somehow an amicable society will emerge. It doesn't always work out—especially when the group is top-heavy with veteran scientists with strongly developed opinions and plenty of ego. I was well aware that in his controversial book *Bones: Ancient Men and Modern Myths,* Lew Binford had sunk his teeth into George Frison's influential interpretations of the behavior of early inhabitants of the New World, based on an important site in Wyoming. Now the two senior archeologists were sharing the same shower bag and latrine.

As it turned out, I needn't have worried on that score at all.

"I made some stabs in the wrong direction, and Lew set me straight," Frison told me when we had a chance to talk. "There's no hard feelings. If you can't take some heat, you're in the wrong business."

I wish more people in academic fields took that attitude—but

George Frison is not a typical academic. He spent his early adulthood as a rancher, entering the University of Wyoming as a freshman at the age of thirty-seven. George finished his undergraduate training in two years; three years later the Anthropology Department at the university had granted him a Ph.D. The following year, at the age of forty-three, he had become chairman of the department. He and his wife, June, are straightforward, friendly, salt-of-the-earth types, sparing of words. I liked them immediately.

I liked Lew Binford too, and his wife, Nancy Stone, turned out to be charming. But one could hardly call him sparing of words. A big, silver-bearded, barrel-chested man, Binford is one of those people whose brain always seems to be working at full throttle. Ideas flow out of him in a stream that at times he seems at a loss to control. It's worth taking the plunge and following him along. Lew is the most influential *thinker* in modern archeology—a prolific iconoclast whose insistence on rigorous self-questioning in the interpretation of sites has opened up a whole new approach to understanding the ancient past. In the process, he had probably set some sort of record for "Most Colleagues Offended in a Single Career." (Given time, Tim White may snatch that trophy away from him.)

The three African students in the group had been hand-picked for their talents by governments keen on developing national research prerogatives. Prosper Ndessokia and Paul Manega were being trained at American universities to assume key roles in Tanzania in, respectively, paleontology and geology. Berhane Asfaw was well on his way to his Ph.D. in paleoanthropology at Berkeley, and would be in an excellent position to help coordinate research in Ethiopia when the time came to push farther into the rich deposits of the Afar and Awash regions. Meanwhile Gen Suwa, already an expert on hominid dentition, was also near the end of his graduate work. He would return to Kyoto University in Japan, remaining active in East African sites.

These students were mature, committed professionals. They were not immune, however, to the flush of excitement that comes when you first arrive in the field. Suwa, finally given a taste of fieldwork after six years buried in the lab, was thoroughly infected with the bug, darting around the Olduvai outcrops as if human fossils were calling to him from under every rock. It seemed that whenever Tim and I sat down to relax after a stint of surveying, Gen would appear in his inevitable blue windbreaker, rock hammer in hand, practically panting to go back out and find an australopithecine skull.

"He's got a bad case of hominid fever," Tim laughed. "Let's hope we all catch it." You have to be optimistic in this business—disappointments come much too often.

While in Berkeley, Prosper Ndessokia had befriended a young Californian named Jeremy Paul and had invited him to spend a few days with us at the Olduvai camp. Jeremy was on the first leg of a round-the-world trek with an ambitiously rugged itinerary. He had outfitted himself with an array of state-of-the-art gear, including a super-lightweight water purifier and a Walkman tape player apparently built to withstand a downpour in Rwanda or a fall from a cliff in Nepal. Jeremy was friendly and ready to help around camp with anything that needed doing, but I could tell that his cool California mannerisms grated on some members of the staff. He had saved up the money for his journey by working in his family's hot-tub business in Marin County.

"Johanson takes over Olduvai Gorge, and one of the first experts he brings in is a Jacuzzi salesman," Tim White kidded me one evening. "Just wait until the boys in the Nairobi Museum get hold of that!"

We had assembled a mixed group, and not surprisingly it was taking a little time for everyone to get accustomed to the others. As it turned out, the initial bonds were cemented with peanut butter. We had all the ingredients needed: plenty of peanuts in camp, my blender from Berkeley, and one working windmill to provide electric current. The peanuts, however, first had to be shelled and divested of their skins. This job required a considerable labor force. On the previous afternoon, everyone had pitched into the business of skinning peanuts. After about thirty minutes we had accumulated enough raw material for several blenderfuls of butter, and a less definable amount of mutual goodwill. We revved up the Waring and soon had a pint of lumpy goo. Served on slices of freshly baked bread, it more than rewarded the effort of carting the machine from the other side of the world.

Halfway into a second batch, the blender abruptly and mysteriously stopped. Gerry Eck took it apart and tinkered with it for a while, growing more involved the longer the blender resisted his mechanical talents. When Gerry took a break, Tim slipped in and took his place, with several others kibitzing over his shoulder. Ten minutes later he threw up his hands and walked off to take a shower.

"*I've* got something that will fix it," Jeremy Paul declared, rooting through his backpack. I half expected him to pull out a featherweight voltmeter and a package of dehydrated blender parts. He reemerged instead with a six-inch length of wire, which he used to bypass the on-off switch. But there was still no sound from the blender. Gerry returned and poked at it for a while, then George Frison had a go, but they only succeeded in further degutting the machine.

Now, a day later, my blender lay strewn about the far end of the table like the remains of a lion kill that had been picked clean by scavengers. Every once in a while somebody would return to it with renewed interest, but quickly the will to tinker would fade. We never did get it to work again, nor did we discover why it had so mysteriously died in the first place.

Bob Walter, the expedition geologist, showed no interest in small-appliance mechanics. I knew Bob well from the Hadar times. A dapper fellow with wide innocent eyes and a wry sense of humor, he always comes in for a lot of kidding in camp, though I'm at a loss to explain why. Perhaps it's a quality he has of seeming perpetually clean, inside and out. After a morning of surveying in some sun-baked gully, Bob would return looking eager and perfectly scrubbed, as if the dust that settled on the rest of us was repelled by some special property of his skin.

Not long after dawn on the first morning in camp, Bob had taken the whole group out to look at the geology of the Second Fault—a five-mile ride along the northern rim of the Gorge. The lorry came to an abrupt stop just a few feet from the edge. A stiff wind pushed against our backs as we walked to the precipice. For a few moments we stared out across the emptiness at the bare, muscular horizons on the opposite side, plunging 250 feet down to the Gorge floor. The fault line was clear: a diagonal gouge running bottom left to top right. To the right, west of the fault, the deposits lay fairly undisturbed, but to the east the entire landmass had sunk down, collapsing the strata into the fault line. Looking down the length of the Gorge, I could see where it narrowed before leveling off into the Olbalbal Depression, a shallow drainage basin lying beneath rose-tinted highlands to the east.

Bob found a little promontory where he could stand in front of us and use the Gorge as an illustration of itself.

"You can get a pretty good perspective here on the entire geological sequence in the Gorge," he said. "Do you see that black stuff at the foot of the cliff and down on the floor? That's the basalt that forms the base of the Olduvai beds. If you walked down to the bottom on this side you'd find almost a hundred feet of it exposed in the cliff face. We know from the original potassium-argon dating done here in the later fifties that the basalt is about two million years old. Above the black lava you've got the sequence of ash deposits that we call Bed I. There's close to a hundred feet of it too, here at the Second Fault, representing some three hundred thousand years. Bed I is composed of volcanic tuffs interspersed with clay stone deposits. Some of the tuffs have very distinctive features, which have been a big help in correlating strata horizontally over the length of the Gorge. Take for instance Tuff IB—that's the band of lighter gray just above the base of the cliff on the right."

Bob continued to point out the successive strata visible across the Gorge, almost shouting to make himself heard over the wind. Bed II, he explained, was a series of clay and sandstone tuffs and deposits, about eighty feet thick at the Fault. It represented some 600,000 years of prehistoric time, and like Bed I had proved rich in fossilized bone and artifacts. These included cranial remains of both *Homo habilis* and *Homo erectus*, along with several extremely important archeological sites.

Lying atop Bed II were the crisply defined sedimentary deposits of Bed III: a rich copper color sandwiched between duller buffs and grays. While certainly the most vibrantly colored stratum at Olduvai, Bed III has yielded few fossils. Bed IV, a brownish-gray series of tuffs visible here only on the collapsed, eastern side of the Fault, has proved similarly sparse in preserved remains. Above Bed IV lies a series of younger ash deposits: the Masek and Ndutu beds, named after lakes at the head of the Gorge, and a bed called the Naisiusiu (pronounced "Nay-soo-soo"), which is the Maasai word for the sound of dry grass rustling in the wind. A lot of local names are Maasai in origin, including "Olduvai" itself, a corruption of the Maasai word *Oldupai,* which means "Place of the Wild Sisal." Found all over the Gorge, the formidable-looking sisal plant grows in clusters of thick tapering stalks, sharp as spears.

Earlier in this book I've referred to the Olduvai stratigraphy as a "layer cake," as if each of these beds was neatly stacked on the one below. As Bob Walter would quickly point out, two million years of geologic upheaval and reworking by the elements produce a much more complex concoction than the cake metaphor really implies.

Consider, for instance, the cake "pan" of hard black basalt underlying Bed I, belched out as molten lava from fissures in the earth's crust. When the lava cooled, it formed an undulating, uneven surface, so that the "batter" of hot ash and detritus thrown down by the volcanoes collected unevenly, filling up depressions in the pan, but only thinly coating the higher ridges. As more layers began to collect, the unfinished cake was further disturbed by violent shifts in the earth's crust—the ravaged strata at the Second Fault forever bearing witness to one of them.

After five major faults and innumerable smaller ones, the layers of the cake would no longer match up horizontally. Meanwhile, the surface of the landscape was being continually stirred and confused by the action of streams, wind, and other natural forces. Next, this disheveled dessert was eaten into by erosion—the cutting of the Gorge itself, beginning some half a million years ago. Only after the older deposits were well bitten into was the "icing" of the cake laid down: more windblown volcanic eruptions. forming the Ndutu and Naisiusiu beds. Here at the Second Fault the stratigraphy might look tidily cakelike, but if we were to get back in the lorry and drive to some other overlook, the geological sequence would be more obscure, and certainly different. It is this horizontal variation running across the Gorge's horizons that has really tested the geologists' patience.

Most of that patience—twenty years of it—was supplied by Richard Hay, now at the University of Illinois. A quiet, unflappable geological detective, Hay walked out, picked at, sifted through, and analyzed the composition of rock in dozens of cliff faces and gullies. His work incorporated a wealth of information to help him correlate the various deposits. By thc time he was finished, Dick Hay had not only reconstructed the geological history of the place, but had synthesized that data into a remarkably convincing portrait of the changing Olduvai environment.

Hay found that during the deposition of Bed I and the lower part of Bed II (from about 1.9 to 1.6 million years ago), the region had been dominated by a wide, shallow alkaline lake that lay at the foot of volcanic highlands to the east. Numerous freshwater streams and small rivers flowed *into* the lake, but it had no outlet; consequently the water alternately covered and exposed a wide floodplain. Fed by the intermittent inundation of mineral-rich volcanic ash, the highly alkaline lake could support an abundance of microscopic algae, which in turn supported fish and birds, themselves prey to larger creatures like crocodiles. Plant life in the lake margin also lured a wealth of herbivores—ancient elephant and their extinct cousin, *Deinotherium*, as well as wild horse, rhino, hippo, and a plethora of bovids. The irony of this Rousseauesque vision of bio-abundance is that periodically it would all be chased out by a new blanket of the very ash that had sparked such fertility in the first place.

One of the most informative clues to the ancient ecology of Olduvai is the wealth of fauna preserved in its piled layers of time. A little over a million and a half years ago, the "profile" of the typical Olduvai inhabitant changed. Animals that lived exclusively in swampy habitats, such as hippo, abruptly disappeared, to be replaced by a greater frequency of species adapted to the drier conditions of a savannah habitat. This "faunal break," echoed in the kinds of pollen preserved and the record of windblown and water-worked particles forming the deposits, coincided with the beginning of widespread faulting in the region.

Perhaps aided by global changes in climate, the giant shifts in the earth shrank the lake to a third of its size and set in motion a gradual eastward march of its drainage sump, until it reached its present location in the Olbalbal Basin some 400,000 years ago. The changing portrait of water accumulation and drainage was of course of more than academic interest: Almost all animals, hominids included, need sources of fresh water. Locate the prevailing watercourses, and that's where you'll find bones and archeological sites. In his work with Mary Leakey, Dick Hay found that by Bed III and Bed IV times, the traces of prehistoric humanity had spread out and dispersed along stream beds.

Today, Olduvai Gorge cuts a twenty-five-mile-long Y into the Serengeti. Where the narrower southern arm of the Y—the Side Gorge—meets the Main Gorge, there is a broad, flat area called the Junction. Before getting down to work, we decided to take the students and other newcomers to an overlook commanding a view of this prime fossil area. The lorry barreled through the Junction and labored up the slope on the other side. After a few minutes we pulled up beside the little museum and grass-roofed *banda* that the Tanzanian Department of antiquities provides at the overlook. I took my camera and wide-angle lens and walked away from the others, vaguely seeking a well-composed shot. Near my vantage point, a grand, isolated rock prominence called the Castle rose up almost to rim level, red with the iron-rich silt of Bed III, plunging in a skirt of gray rubble toward the Gorge floor.

What makes Olduvai such a rich fossil site is its unique geological history. Two million years ago, the wide, shallow alkaline lake at Olduvai drew an abundance of life to its shores. The bones of many animals, hominids included, were buried by sediments and volcanic eruptions. Within the last five hundred thousand years, major earth movements created a depression to the east, and the flowing waters of a stream began to cut through the accumulated deposits. Today the Gorge is one hundred meters deep, exposing the fossilized bones buried when the lake still dominated the terrain.

Two hundred feet below, the Junction lay like a tableau, every contour of rock or flush of vegetation crisply outlined by the oblique light of the morning. The dry bed of the Olduvai River wormed its way through a broken patchwork of aloe, acacia, and wild sisal. The scene possessed an inanimate beauty, infinitely receding, like a view through the wrong end of a telescope. I put the camera aside and gazed down, almost hypnotized. Off to my left I could hear Lew Binford and Bob Walter discussing some geological points, but their voices too sounded distant, as if I were caught suspended between the present moment and the unreachable timelessness of the Gorge at my feet.

Suddenly two vivid images seemed to merge in my mind and impose themselves on the scene below. The first was the view of Lake Manyara, where we had stopped the day before. Almost two million years ago, the eastern margin of the similar Olduvai lake lay just where I was looking. Now in my mind I could see the ancient water, its shallow, sun-dappled surface choked with reeds. The cliffs on the opposite side of the Gorge disappeared; I was no longer standing high above the lake, but instead on a gentle rise beside it. I imagined it so clearly that I could almost smell its saline pungency, a mix of brine and rot: Hippo and crocodile basked along the banks, and a gaudy ribbon of flamingoes fringed the water's edge.

The second image was more disturbing. It leaped into my conciousness from the dream I had experienced that restless night in Dar es Salaam. There was a lake in that dream too, brightly colored and inviting, but naggingly out of reach. I wanted to immerse myself in its waters, but something kept holding me back. As in the dream, the air now was full of a insistent murmuring. I looked back toward the museum overlook: Tim, Bob, and the others had all disappeared. In their place, a party of Italian tourists had emerged from a trio of safari combis. They were talking all at once and pointing into the Gorge. I glanced down. Tim and Bob were slowly picking their way across the slope toward the Castle. Prosper and Gen followed behind, and I could make out Lew Binford and his wife, Nancy, farther down. Just then Tim looked back in my direction.

"Hey, Johanson!" he shouted. "You come here to find hominids or to take pictures? Get your butt down here!"

I snapped out of my trance. Buckling the camera strap around my waist, I carefully dropped down onto the slope.

• • •

We spent the first day surveying around an area just to the east of the Castle called HWK (Henrietta Wilfrida Karongo) a fossil site defined by the slopes of a dry stream bed. It was at HWK the summer before that we had cleared off the five-square-meter patch of slope. When I caught up to Tim he was squatting on the ground next to the patch, with Gen and Berhane standing beside him. The boundaries of our 1985 square were still clear, but they had filled up with a jumble of debris.

"What do you think, Don?" said Tim. I looked down at the square. Even with a quick scan I could see a piece of horse limb, some tortoiseshell, and a couple of fossil teeth, as well as a beautiful pink-quartz chopper and some other stone tools.

"I think Olduvai must be the liveliest dead site in East Africa," I said. "It will take us a half a day just to inventory this stuff."

"Keep in mind that as far as we can see at first glance, none of this material is *in situ,*" Tim said to his students, meaning that the material was lying on the surface instead of eroding directly out of the deposit on which it was found. "It was all carried down by the rain this spring, from God knows where up the slope. If I've got a gripe about Olduvai, that's it—there's just too much *time* piled up in these deposits. If you don't find your fossil *in situ,* it could be two hundred thousand years old, or two million."

"True enough," I said, "but the purpose of this experiment was to demonstrate that erosion constantly exposes new material, not to ascertain where that material comes from. And I think we've made our point."

"Not until we've got this stuff all plotted and catalogued," Tim answered. He turned one of his sly, lopsided grins on his two students. "Either of you guys want to volunteer for that job, while Johanson and I go on ahead? I got a feeling we're going to find Zinj's girlfriend in this damn gully."

Gen and Berhane knew their teacher well enough not to bite on the tease. He wouldn't ask them to spend the first day in Olduvai Gorge counting rocks and anonymous bits of bone.

"Zinj's girlfriend?" said Gen. "HWK is half a million years above the Zinj floor."

"So, maybe she likes older men," Berhane added softly. We all laughed and started down the stream bed, leaving the work on the square for later.

The four of us fanned out across the western slope above the stream bed, which was heavily carved by rain runoff—a good surface to explore for newly exposed material. I found a deep gully and began to work my way up, eyes on the ground. The light was perfect—plenty of midmorning shadow to add definition to the groundscape.

Still, I knew that the chances of finding anything unusual were slim. Discoveries are rarely made the first few days in the field. It takes time to attune your eyes to a radically different scale. In the landscapes of our normal routines, we're used to scanning for big things—our car in the parking lot, or the gas station on the corner where the directions say we should turn left. And the backgrounds we see in our everyday life are usually simple—pavements, carpets, painted walls. Out in the field, you're asking your eyes to scrutinize a clutter of tiny objects of subtle textures and color, all randomly arrayed. HWK is a fairly rich site, and I could easily make out the larger fragments of bone lying here and there. But these big pieces were the ones that had probably been picked up, evaluated, and discarded by a dozen other fossil hunters who had walked this slope before us. Of course I checked each one again, just to be sure. But I was well aware that at the same time my eyes might be blind to smaller, subtler clues—the faintest sheen cast by bone eroding out of a rain-cut hollow, or an unusual pattern of relief in the dust.

A few yards to my right I noticed a hand ax lying on a little ledge. No hand axes have been found *in situ* at HWK, and this one had no doubt slid down the slope from some level up above. It had little value, therefore, as an archeological artifact. I traversed the pebbly lag anyway and picked up the tool. It was a classic example of the type: a symmetrical teardrop of worked quartz about six inches in length, with flakes knocked off on two faces to form an edge. The broad end fit comfortably in my grip. There's a strange urgency in these ancient axes: As soon as you pick one up its balanced heft seems to compel your arm to *do* something with it, if only to rock your elbow and appraise its weight in your palm. The "biface" hand ax typifies a period called the Acheulean, which endures in the archeological record for a stunning length of time. Tools very much like the one I now held had been found in European sites as young as 500,000 years, as well as in Olduvai deposits dated at a million and a half.

Whatever they were used for, clearly the hand axes and other Acheulean tools were doing it efficiently. Compared to the longevity of the hand ax, the invention of the automobile—or for that matter the wheel itself—strikes me as a sort of cultural whimsy, a fleeting bit of gadgetry. Thousands of generations separated me from the individual who had knocked this flake of rock off a boulder and fashioned its shape. But I, a user of garage-door openers, power saws, and electric blenders, could instantly recognize it as a *tool*, potent with human purpose. Gently, idly, I tapped the blunted point against my palm. For a moment a few thousand generations didn't seem like much time at all. I could count them off in less than an hour with gentle taps of the tool in the palm of my hand—parent to child, parent to child. There was something very reassuring, almost liberating about that thought. I smiled to myself and put the hand ax back on the ledge where I had found it.

I've already mentioned the first stone tool recorded from Olduvai—the one Louis Leakey brought back to camp in 1931 to win his bet with Hans Reck. By the time Mary Leakey retired from the field in 1983, the Olduvai horizons had disgorged over 37,000 more tools—certainly the most grandly informative record of early man from any one site on earth. In addition to their sheer quantity, the Olduvai artifacts are of remarkable quality. Most important, the nature of the tools changes through time, bearing testimony to the increasing sophistication of human culture through its first million and a half years.

The earliest stone tools found in Olduvai Bed I sites date from very close to the beginnings of the archeological record. In Mary's evaluation, the "tool kit" used by the hominids in the very oldest sites, such as the Zinj floor at 1.8 million years, consisted of rough choppers, flakes, and similarly unembellished forms, all created by a few strokes of a hammerstone on one side of a rock core. The forms of these "Oldowan" tools became more complex the higher up she looked in the geological strata—rough-faceted polyhedrons, for example, giving way through time to remarkably smooth spheroids. Twenty years earlier, Louis had believed that this gradual trend toward sophistication culminated in the elegant Acheulean tradition. But Mary could see that in fact the hand axes and other Acheulean tools appeared quite abruptly in the middle Bed II, about 1.6 million years ago. For the next million years, the Oldowan and Acheulean industries continued on side by side, all the way into upper Bed IV—*but never occurred at the same site.* This odd duality led Mary and her colleagues to suggest that perhaps the Acheulean tools represented the culture of an entirely different group of hominids, one that had arrived at Olduvai from some other region. That theory was made all the more convincing by the sudden appearance of *Homo erectus* fossils in the Gorge at about the same time.

Over the years, Mary and her colleagues chiseled from the archeological record of the Gorge a carefully worked portrait of early human activity. At its center was the increasing importance of hunting and the role it played in establishing uniquely human patterns of behavior and social organization. The "hunting hypothesis" was no idle theory: It was bolted to a structure of evidence so broad-shouldered and powerful that no one even considered doubting its validity.

No one, that is, until Lewis Binford came along. Binford's shoulders are pretty broad too, and he plays the game of archeology much the way a good defensive end plays football—with quick intelligence and murderous elan. In his *Bones* book, Binford came down hard on the conventional interpretation of the Olduvai archeological record, questioning the very foundations of the hunting hypothesis and upsetting a lot of people in the process. I will save the substance of that debate for later. Let it suffice to say now that I wasn't sure that I agreed with everything Binford said, but I knew his ideas were worth taking very seriously—enough to invite him to Olduvai.

The evening after that first surveying excursion at HWK, he and George Frison were clearly excited.

"We found tons of evidence of hominid behavior out there," Lew told me after dinner had been cleared away. "There were bones showing cut marks and impact fractures all over the place."

Cut marks are the preserved scratchings from hominid tools found on faunal remains; impact fractures are breaks in bones indicating by their pattern that they were made by the crushing blow of a stone implement. Though he had never visited Olduvai before, Binford knew its topography as well as his own backyard in Corrales, New Mexico. Now he sketched a quick map on the back of an envelope lying on the table.

"The way we see it, there are really two places to look for good sites in the Gorge," he said. "We know that the hominids had to rely on water sources like everything else out there. In the dry season, the place to be would have been over here toward the west, at the Ndutu Stream entrance to the lake basin. In the wet season, you've got drainage from those volcanoes to the east into *this* region." He circled on his map an area around the Castle, including

HWK. "If you could excavate sites in both these areas, you might begin to see the difference between dry- and wet-season usage by hominids. Now *that* would be something worth doing."

I listened with interest. Not many people had done comparative studies of seasonal use by early hominids. In different seasons, the hominids' diet would change—and if their food sources varied, so might the kind and pattern of tools and discarded bones they left behind. Millions of years later, the archeological record would be fundamentally different too. This was just the sort of insight I was hoping for. We talked for a while more around the table, then went to bed.

After a late start, the second day in camp was devoted to surveying in a couple of Bed II sites in the Side Gorge. At a place called FC, Tim stopped to give his students an on-location lecture on field practice. I was fifty feet downslope, but his deep, melodious voice carried well.

"Let's pretend this chunk of equid femur is a hominid," Tim said, picking up a fragment of horse leg bone from the surface. "Is it *in situ*? No. Suwa—any matrix on that bone?"

Tim handed the fossil over to Gen, who shook his head and handed it back. "Matrix" refers to any mineral deposits clinging to the bone that might help identify which stratum in the slope it originally came from. If the fossil is not *in situ,* matrix is the only sure way of pinpointing its age.

"Right. The first thing you'd do now is check the thing for fresh breaks. You can see by the sharp edges that this bone has broken up only recently, so more of it might still be around. So the second thing you'd do is get the hell off this slope before you step on anything. You'd want to systematically put some kind of square around this area to confine your search area. Cut that square into smaller segments, then start at the bottom of the slope and work up, picking every bone fragment out of each segment. Only after the surface is completely checked do you think about digging.

"Let's say you've cleaned the slope and still haven't found anything else," Tim continued. "If you decide to dig, you're looking at a couple of weeks of work. Minimum. So you've got to ask yourself—for Christ's sake, is this worth the time? If what you've got in hand is a five-million-year-old hominid femur, you bet it is. But if it's a chunk of *Homo erectus* limb from Bed III, you might say to hell with it. In the field, you're always running against the clock, and those

The stone circle at DK in Olduvai Gorge, thought by Mary Leakey and her associates to represent the remains of an ancient dwelling

decisions have to be made. Every minute you spend digging here is a minute lost somewhere else in your site."

Tim threw the stand-in hominid bone back on the ground. "Now let's take a look over here," he said. "Asfaw—what can you tell me about this tuff?"

The three of them climbed into the next gully over, out of earshot. I continued on up the slope.

With the students in mind as much as the visiting archeologists, we had planned for today an excursion to a famous site called DK out toward the Second Fault. In 1962, Mary and her team had uncovered a remarkable feature on what appeared—judging from the tools and broken faunal bones scattered about—to be a home base of early hominids. While clearing away some blocks of lava, Mary noticed that the remaining blocks seemed to form part of a rough circle about thirteen feet in diameter. Some of the stones were even piled up on each other. The pattern prompted Helson Mukuri, a long-time member of the Leakeys' African staff, to point out that he knew of nomadic tribes in parts of Africa who built shelters out of branches and animal sinks piled on stone bases. Could this "stone circle" be the remains of some prehistoric structure? Mary replaced the lava blocks that she had already removed. Inside the circle, the concentration of broken faunal bones was much less dense than outside, suggesting that the resident hominids had tossed away the discarded remains of their meals. Most astonishing of all was the age of the site: The lava blocks rested almost directly on the black basalt at the base of Bed I. Meaning that nearly two million years ago, ancestors of man were constructing and living in rudimentary houses.

Not surprisingly, the stone circle at DK lent new support to belief in the chronological depth of humanity. Through the sixties and seventies, as other archeological sites at Olduvai were excavated, the interpretation of the circle became accepted as fact. "DK . . . demonstrates beyond most reasonable doubt," states an archeology textbook published as recently as 1984, "that early men were already building substantial shelters. . . . Its excavation must rank as one of the most important of all Stone Age discoveries."

Perhaps the writer of that text considered Lew Binford an unreasonable doubter. In Lew's opinion, the DK stone circle was a perfect example of an artifact that had much more to do with modern myths than with ancient men. In the *Bones* book he pointed out that in an area of about 250 yards around the site, the excavators had found nearly 4,600 crocodile teeth shed by juveniles. Obviously, the deposit in which the site rested took a considerable period of time to accumulate all those teeth—perhaps thousands of years. Wasn't it jumping to conclusions to call the site a "home base" in the first place, when so much time was represented? The faunal remains around the stone circle, in fact, were just what one would expect to find at any location near a lake margin, where many animals would have lived and died naturally through generations, without their carcasses having been transported and consumed by hominids. Yes, there were tools on the site—but there are tools almost everywhere in that region of the Gorge, and these were not arranged in any meaningful pattern in relation to the stone circle.

Binford conceded that the circle itself was indeed "enigmatic," but *unless one had already concluded that DK was a home base of hominids,* there was no reason to think that it was deliberately constructed. It might just as well be a random association of rocks laid down by some natural agent—for instance, a peculiar current in the flood margins of the lake, or the action of growing tree roots. Having established the notion of a shelter, however, the excavators then turned around and used that interpretation to confirm their belief in the site as a living floor—precisely the observation that had given rise to the shelter idea in the first place. For Lew Binford, at least, there was nothing circular at DK except Mary Leakey's reasoning.

Toward 8:00 A.M., the lorry dropped us off at the rim of the Gorge above DK. For years the site has been protected from the elements by a sturdy, tin-roofed stone building, erected with funds donated by Gordon Hanes of the Hanes clothing company. We climbed down toward the little building a few hundred yards below, poking around in the deposits along the way. We made it to the bottom of the Gorge by ten. It was hot and windless, the rubble all around shimmering in a baked stillness. When we reached DK, Binford edged ambivalently around the flanks of the site before going in, like an enemy pilot come to pay a surreptitious visit to a city that he himself had reduced to ruins.

Inside the stone building the air was much cooler. As many of us as would fit filed into a narrow walkway in the middle of the room. The place had a solemn, cloistered feel to it, like a secluded chapel. A pale light filtered in from above, so colorless and weak that it barely seemed to reach the floor. For a moment we leaned over the wooden railings of the walkway, contemplating the bones and artifacts strewn on either side. The stone circle was on the left. At first glance it looked like little more than a pile of rubble, loosely piled into small heaps along one side.

"So what do you see?" asked Lew after a while.

"I see a pile of rubble piled up into small heaps on one side," I said. "Though if you look at it a little closer, and keep in mind Mary's plan of the stone circle in her book, I can make out a ringlike shape to it."

"That brings up a couple of points," said Lew. "First of all there's Mary's plan. As Tim has pointed out, it just doesn't correspond to what's here. I'm not saying she drew the plan carelessly. But don't forget that she removed a lot of the lava blocks before she realized that they belonged to the alleged circle. Without intending to, her inclination to see a circle could have influenced the way she replaced the blocks."

"What's the other point?" asked Bob Walter.

"Don said it himself—if you take a *closer look* you begin to see a circular shape. But on that closer inspection, your eyes are *looking* for a circle, so naturally they're inclined to see one. The really odd thing about this site is that you can find circles all over the place, if you look for them—there's one to the right over there, and another one just behind it.

"I think I can see them," said Gen Suwa. "But those circles are a lot smaller—only a couple of feet in diameter."

"Exactly," said Lew. "So what do you think they are? Hominid dollhouses?"

We filed back into the bright heat outside and surveyed for a couple of hours in a series of gullies near the site. There were plenty of croc teeth and some root casts—fossilized plant debris—lying around to bear witness to the marshy lake margin that once lay upon the site. But mammal bones were sparse, especially those showing hominid activity. As the sun rose higher it became increasingly hard to make out anything at all in the deposits. We headed back out of the Gorge, and a little while later I could hear the lorry laboring along the rim, to meet us in time to take us back to camp for lunch. The last thirty meters of slope were tough going, and I was pretty well beat by the time I made it to the rim.

Bob Walter was there ahead of me, smiling, dangling his legs over the edge, and looking as fresh as an off-duty lifeguard sitting by a swimming pool. I flopped down beside him. Below, I could see Lew struggling up the slope, burdened by some video equipment strapped over his back. When his head poked over the rim, the blue bandanna he wore as a headband was drenched in sweat. The three of us stared back in silence down the way we had come. At the bottom of the Gorge, the modern stone building covering the DK site was plainly visible, charmingly incongruous against the raw African landscape.

"I don't care what you say, Lew," Bob said, "that sure looks like a house to me."

We were greeted back at camp by a lunch of cabbage soup, followed by bread served with a gelatinous canned meat the color of aging flesh. I can be pretty sure of the menu, because lunch that summer at Olduvai was nearly *always* cabbage soup, bread, and canned meat. We added plenty of Tabasco sauce and washed it all down with pitcherfuls of raspberry Kool-Aid. The morning's surveying had exhausted everyone and there was not a lot of talk around the table. After lunch, most of the crew retired to rest through the hot part of the afternoon. I decided to stay in the open central room and record the events of the last three days in my journal.

It didn't take me long to realize that there weren't many events to record. Apart from introducing the students to the site, what had we accomplished so far? Systematic surveying had yet to begin. Binford was getting a bead on the archeological possibilities, and he and George Frison had made some tantalizing suggestions about how best to proceed. Their ideas were ambitious, involving large-scale excavations of sites over a wide area of ancient landscape. But that sort of work would require huge expenditures. Where would the money come from? What good were ambitious ideas, when we were hanging on in the Gorge by a shoestring as it was, through the good graces of some private donors? I suspected, furthermore, that when Fidel Masao arrived—and it was Masao who was in charge of archeology at the Gorge—he would be inclined toward a more conservative approach. I put the journal aside and stared out at the weaver birds flitting about in the acacia.

I was anxious for Masao's arrival. I hoped he would bring with him the promised permit to work at Laetoli—as well as the means to get there. Olduvai might be a good bet in the long term, but Tim and I both knew that we'd be better off spending what little time we had this summer surveying in the richer, older deposits at Laetoli instead. Without grant support, however, we were relying on vehicles loaned by the Tanzanian Department of Antiquities. The lorry was needed back in Arusha the following day, which would leave us with a single Land-Rover. It would not be safe to take it to Laetoli and leave those behind without transportation. Laetoli was only twenty miles away, but unless we could put our hands on another vehicle, it might as well have been in South America.

Gerry Eck came around the side of the building, startling the flock of weavers out of the tree. A moment later I saw them descend into another acacia down by the camp gate, catching the sun like a handful of glitter tossed into the air. Gerry walked out to the middle of the yard, where a defunct anemometer—a device to measure wind volume—was mounted on a post. Having given up on my blender, Gerry had decided to take on the challenge of bringing the instrument back to life. For some reason his relentless industry irritated me.

"Look's like Eck is keeping himself busy." I hadn't heard Tim come up the path; he was sitting on the low stone wall in front of me.

"Great," I said. "Any day now we'll begin to know again how much wind is blowing out across the Serengeti."

"You sound a little sour."

"Sorry. It's not Gerry, of course. It's just that I'm beginning to wonder how much we can accomplish here in four weeks. And what happens if we don't accomplish anything.

"I can't say those thoughts haven't crossed my mind too," said Tim.

"There's something about this whole thing that doesn't feel right,"

I said. "I shouldn't complain—the climate's good, we've got clean quarters and enough food, and in the morning somebody whisks away my dirty laundry and returns it washed and folded before lunch. But this is supposed to be an *expedition*, for God's sake, not a vacation. In the Afar, we slept in tents, worked in hundred-and-ten-degree heat and kept one eye out all the time for bandits. Here, the most threatening intruders we have to fear are some overenthusiastic tourists poking around the gate."

We stared out a while in silence. Then Berhane and Gen came around from the other side of the building. They had their rock hammers and canteens.

"We're going down to the Junction to survey," Berhane said. They started to walk off toward the gate.

"Hold on a minute," Tim called after them. "I'd better go with you guys and make sure you don't get eaten by a lion or something." I could tell that he didn't want to go. But it was typical of Tim to use every opportunity to teach his students.

"What about you, Don?" he said to me. "You're not going to find any hominids by moping around up here."

"I might as well keep you company," I said. Tim was right. There was no point in brooding.

Gen and Berhane waited while we fetched our equipment, then the four of us trudged down the road toward the Junction. We were heading for a promising area that Tim had noticed the year before, a low plateau of deposits just to the west of the Castle. In jest, we named the place "TWK," short for Tim White Karongo. There was plenty of stuff visible *in situ*, well-preserved bone with a pretty rosy tint to it. As usual Tim kept his students busy, throwing them bone fragments to identify, pointing out tuffs, explaining procedures as we went along. In one crevice we found a bovid mandible eroding out, broken up into fragments held together only by the pressure of the deposit surrounding it.

"This fossil will be lost next year to erosion," Tim explained. "If you wanted to collect it, you could do one of two things. Either take it out in pieces, put them in a bag, label it and take it back to camp. Then see what you can put together later in the lab. The alternative is to bring some Glyptal or other preservative in here now, and let it penetrate that sucker, freeze the whole thing. Let it dry, then dig away a little and pour in some more Glyptal. Then dig some more, and so on. The point is to take it real slow. It would take a day or two to get it out of here in one block. People make the mistake of putting preservative on the thing, then they try to lift it out two hours later. Take the time to do it right."

We surveyed for a couple of hours around TWK and the base of the Castle. By five o'clock the light was beginning to fade, and we started back for camp. The sun was just settling down toward the rim of the Gorge when we reached the place where the road started to climb back up to the rim.

"There's still twenty minutes of light left," said Gen.

"Let's follow the road across the Junction toward FLK, then climb up from there," said Tim.

It was one of those weightless little decisions that only later, in hindsight, explode out of obscurity. We crossed a flat area, walking along parallel to the road, eyes on the ground. Slowly the Junction filled with dusk. On the slope of a low hill I saw Tim bend down to pick up what looked like a limb shaft of some kind. As he bent over, something else caught his eye—a cylinder of dark-brown bone an inch long. He held it up close to his face and squinted through his thick glasses.

"Damn! This is *hominid,*" he said, almost in a whisper. I looked over his shoulder. In his hand was the elbow end of a primate ulna, or forearm. It was small enough to be that of a baboon but lacked the strongly projecting flange of bone in back where a baboon's powerful triceps would insert. That could mean only one thing. I reached down and picked up the limb bone that had first drawn Tim's attention.

"This looks like a piece of humerus," I said. It seemed too incredible to believe—pieces of a hominid skeleton, only a few paces from the tourist road to FLK! Carefully, the four of us got down on hands and knees and started searching through the dust.

"Here's a maxilla," Berhane said after a couple of minutes. His tone was so quiet and matter-of-fact that at first I thought he'd simply found some bovid's jaw. But when I looked up at what he held in his hand I felt a surge of joy and relief go through me like a shock. It was a substantial piece of hominid upper jawbone. There were no teeth intact, but some tooth sockets were clearly delineated. We had fallen upon one of the rarest kinds of discoveries: remains of the cranial region of a hominid found along with skeletal parts from below the neck—a far more valuable association than either skull or limb bones alone. Then it struck me: The first piece of Lucy I had found had also been the upper end of an ulna. The coincidence was uncanny. I got back down on my knees and squinted, trying to pick up the outlines of things in the gathering darkness. Just below my left shoe, I picked up a fossil tooth root, the same purplish-brown color of the other hominid pieces. Another half step, and I might have crushed it.

"Hold it," I said. "We've got to leave this till morning. It's getting too dark to see what we're doing."

"Don's right," said Tim. "This baby's been lying out here on the surface for a few decades at least, and it's not going anywhere tonight. We should mark these finds and clear out of here before we bust anything."

We placed piles of stones where we'd found the fossils. Then Berhane, Tim, and I gingerly stepped away. Gen was oblivious, still prowling on his hands and knees, his face tensed with an almost animallike concentration.

"I think Gen's got the worst case of hominid fever I've ever seen," I said.

"Come on, Suwa," Tim laughed. "You can find the skull for us in the morning."

Tim carefully wrapped the bones in his handkerchief and we started back toward camp. As we crested the rim, the last splash of the day's sun greeted us full in the face, a warm, robust light turning the plain to liquid gold. We walked four abreast toward the camp gate. Later, somebody back in camp told me that we looked like four happy warriors coming home from the battle. For dinner that night Gerry put away the Kool-Aid and brought out a case of Tanzanian beer that he'd kept very quiet about. We talked over the bones and made plans for the next day. I turned in early and got out my journal.

"At last, I can sleep easy," I wrote, and sketched out the afternoon's events. After only three days, the old Gorge had returned our faith in its promise. I fell into a dreamless sleep, unaware that our problems were just beginning.

CHAPTER SIX: *Day of the Dik-dik*

There's nothing worse than having a skeleton and you don't know what it is.

—Tim White

NONE OF US HAD THE patience to wait for dawn the next day to begin. We rose early and followed our flashlight beams around in the dark, wrapped in sweaters against the cold. I washed up and joined the others gathering in the central room. We warmed our hands on tip cups of *kahawa*—strong Tanzanian coffee tasting a little of chalk and smoke. Stephan, the cook's assistant, brought out platterfuls of crimson-skinned bananas and papaya strips, eggs scrambled with tomato and purple onion, and fresh bread baked over last night's coals. As we ate, the horizon lifted to the east, the vault of the sky above losing its stars to the infusion of pearly light from the approaching dawn. Then the sun crested Ngorongoro, flattening and spreading out like a flow of magma, as if the ancient volcano were reliving its youthful devastations.

The hominid we had found was still a stranger to us. This was hardly surprising, considering what little material we had in hand. But as scientists, or simply as human beings, we felt an overwhelming urge to probe its identity, find a name for it, pin it down. Leaving aside the unlikely possibility that this hominid would turn out to be an entirely new species, our choices were limited by what had already been discovered in the lower Olduvai beds. We quickly ruled out *Homo erectus,* best represented at Olduvai by a beautifully preserved skullcap found by Louis Leakey in upper Bed II deposits. The ulna Tim had found was far too small to belong to any *erectus* individual known, even if it *had* washed down from upper Bed II deposits—which, given the flat terrain at the site, was extremely unlikely. A more promising "suspect" was the robust species *Australopithecus boisei,* represented by Zinj himself. A second possibility was *Homo habilis,* the first true species of man.

After dinner the night before, we had cleared away the beer bottles and laid the fossils out on the table. We passed the bones from hand to hand, the light from the one dim bulb overhead barely illuminating the faces all around.

"Look at the size of the root sockets of that premolar," said Berhane Asfaw. I sat down next to Berhane to get a better look. This scrap of jaw held no teeth that could give us an easy estimation of molar size. But the ghosts of some teeth were visible—barely—in the bits of preserved tooth sockets.

"It looks like another premolar root coming in there, and possibly a third one behind," I said. "So you've got a premolar that's three-rooted and real broad. A big Zinj-y kind of cheek tooth."

"Those roots might just be flaring out," said Gen. "The tooth crown might be a lot smaller than we think."

"Maybe. And the jaw itself looks real small. What do you make of it, Tim?"

Tim was across the table, preoccupied with the ulna. He looked up and I handed him the maxilla fragment. For a minute or two he turned the fossil back and forth in his fingers, like somebody playing with a Rubik's Cube.

"If I was desperate," he said, "I'd say this is a female robust australopithecine. But I'm not desperate. Not yet."

After breakfast we loaded the lorry with the equipment we would need to sift through the pebbly epidermis of the area around the find. With the exception of the cameras, the paraphernalia of field paleoanthropology hasn't really changed much since Eugène Dubois was combing the river beds of Java for *Pithecanthropus erectus* a hundred years ago. We brought along rock hammers, whisk brooms, some twopenny nails, a ball of surveyor's twine, and the stack of shallow metal pan called *karai* in Swahili. Though we probably wouldn't need it right away, we also brought along a handmade sieve—a two-foot-square wooden tray with a screen bottom, fitted with handles to allow two people to shake the tray back and forth between them. It was all very primitive stuff, and I suppose I can understand why some laboratory scientists—those who would feel naked without a couple of million dollars' worth of technology purring around them—dismiss field research as some kind of romantic anachronism. But until somebody figures out a better method of looking for needles in haystacks, we'll just make have to make do with our wooden sieves and dented karais.

The lorry dropped us off at the new site and headed back to Arusha; from here on in we would have to make do with a single, elderly Land-Rover. I watched the truck climb the road out of the Gorge. A quarter mile away, the copper-waisted Castle presided majestically over the Junction. Just to its right, the thatch-roofed *banda* of the tourist overlook appeared like a bump on the grassy rim of the Gorge. We couldn't have picked a more public place to find a hominid.

Naturally, all the scientists, students, and guests in camp had come along, and naturally, everybody's urge was to rush in and start looking for bones. But the last thing we needed was fourteen pairs of Vibram soles traipsing about mimicking a passing herd of wildebeest. Tim defined a rough no-man's-land around the spot where Berhane had found the maxilla. Then he asked the students to come up front.

"Ndessokia—what's the first thing you want to do in a surface find like this?" Tim asked.

"Try to get some idea of the provenience of the specimen," Prosper answered calmly. "Place it in the stratigraphy as close as we can."

"Okay. And meanwhile, get the site documented and photographed before we mess it up anymore. Now where are we, geologically?"

We all looked around and took stock of the terrain. We were standing on the north slope of a gentle knoll in what was otherwise a flat wash flanking the road. The knoll was barren of vegetation except for the top, which was capped by a cluster of sisal bushes and other ground growth. Just below the bushes, I noticed a circular patch of leathery brown pellets. Dik-diks—shy, tiny African antelopes—habitually return to defecate in the same place. This was one such dik-dik "latrine." A few yards to the southwest, the Gorge wall rose up about fifty feet in a steep, gully-cut series of deposits. I could trace these horizons on for a hundred yards to the south, while to the north, a finger of low, bush-covered sediments blocked the view toward FLK. From the spot where we'd found the bones, a shallow eroded gully ran roughly eastward for about seventy feet to the road, continuing on the other side into a jumble of black lava outcrops and thicker vegetation bordering the dry Olduvai River bed.

"That's basalt down there across the road," Berhane said, "so here we must be pretty near to the bottom of Bed I."

"Right," said Tim. "And since the fossil wasn't found *in situ,* it could have come from anywhere above. There's no matrix on any of the bones we've found either. All you can do is make probability statements."

"The bones are in pretty good shape, so chances are they haven't traveled very far," Gen offered.

"Good thinking. Also consider how fragile a maxilla is, compared to other cranial parts. If that piece of jaw had rolled a long way to where Berhane found it, there probably wouldn't be that much left of it. So we can hope that it weathered out of the real soft stuff underneath this protective lag."

Tim kicked away some surface detritus, then dug the toe of his boot into the crusty surface to expose the softer sediment underneath.

"Maybe there's more in there, maybe not," he said. "All we know for sure is that last night we found three hominid bone fragments in the dark. And I'll be real surprised if we don't find anything more on the surface today. So Asfaw—where do we start looking?"

Berhane looked around the site before answering.

"I'd start back by the road and work up this eroded gully," he said.

"It looks like everybody was paying attention yesterday," Tim said. "We clear the most disturbed possibilities first, and close in on the area of high bone density—which is right here on the northwest shoulder of this rise. So let's get the area roped off and get to work."

While Tim and Gerry oversaw the preparations for the surface search, Bob Walter took off to explore the surrounding horizons for clues to the geological context. Meanwhile Berhane and I took care of photographing and documenting the locations of the individual finds. In the process I turned up a couple of chips of tooth root—promising material, because it indicated more might be buried just

Berhane Asfaw searches surface debris at Dik-dik Hill for signs of the hominid. Behind him are Bob Walter and Tim White.
DONALD JOHANSON

below the surface. Scattered around the area I also noticed a number of crude quartz tools indicative of early Bed I sites—and no hand axes or other Acheulean artifacts that would have rolled down from higher levels. Again, this was the merest suggestion—but perhaps the hominid site was not quite so contaminated by future time as Tim feared. We'd know a little more after the geologist's report. And we'd know a hell of a lot more, I thought, after we found a piece of this hominid *in situ.* I was almost certain that we would—if not today, then tomorrow or the next day.

As if to reward my confidence, Gen Suwa suddenly let out a shout. He'd made his first find—a brownish-black triangle of hominid cranial bone about an inch square, lying on the surface of the knoll just a few yards away. It wasn't *in situ,* and it wasn't an australopithecine skull, but it might be a piece of one. Watching his beaming face, I remembered that just a few days before, we had almost left Gen back in Arusha, cooling his heels until his suitcase turned up. Instead, he had just recognized what could prove to be the first clue to a mystery of overwhelming importance. If enough of the rest of the cranium lay sequestered beneath the surface, we might be able to piece it together and estimate the brain size of this hominid. *That* would be a discovery worth the sacrifice of a few possessions.

Tim walked over and took a look at the bone.

"Congratulations, Suwa." He smiled. "A couple dozen more of these and we might know what the hell it is we're looking for."

Gerry finished laying out a rectangle of surveyor's twine fastened to nails, framing the gully leading down to the road. This would be our work area. Now he and Tim took another length of string and laid it parallel to the road, about a foot inside the rectangle. The surface search would be done in a line on hands and knees, shoulder to shoulder, one foot at a time, moving the string ahead of us. Tim took a central spot to "quarterback" the search, with the students near him. Gerry, Lew Binford, and I found places on the flanks. Completing the line was Mrisho Ramadhani, a staffer at the National Museum who was being trained for fieldwork. A quiet, bespectacled fellow, Ramadhani was known by the nickname "Mzee," a Swahili term of respect meaning "Old Man." Meanwhile, George Frison, who was troubled by a bad back, would remain upright, surveying around the site.

The procedure for a surface search is simple: We scrutinized the patch of dusty gravel immediately before our eyes, putting anything vaguely bonelike in a marked plastic bag. When every chip of bone and splinter of tooth root had been picked out of the narrow channel between ourselves and the movable string, the two men on the ends of the line would get up and shift the string another foot forward. We then crawled forward ourselves and the process repeated itself. Unambiguously hominid remains were to be bagged separately, their location marked with a flag. Later, the distribution of the bones of the surface could tell us a great deal about where the fossil was eroding out. Meanwhile, any stone tools we found were to be left in discrete piles behind us. If we were persistent, by dusk we might reach the spot where Berhane had found the maxilla—a distance of some seventy-five feet. It was backbreaking, knee-scuffing manual labor, growing harder as the sun climbed in the sky. And we were loving every minute it.

"I'd be climbing over the slopes on either side of you guys, and every once in a while shouts and cheers would carry over to me from the site," Bob Walter told me afterward. "I'd know then that you'd found another piece of the hominid. For once I wanted to be a paleoanthropologist."

The first cheer went up shortly after nine, when Lew Binford, working the southern shoulder of the little gully, turned up a tow-inch piece of femur, or thighbone. Limb-bone fragments are especially valuable when they include details of the *ends* of the bones, where they articulate with the neighboring bone to form a joint. Lew's fragment was not one of the articular ends, but it was the next best thing—a two-inch stretch of the femur shaft and neck broken off just where it started to flare toward the proximal end. (In anatomy, *proximal* refers to the part of a limb or other structure nearest to the trunk; the part farthest from the trunk is called *distal.* Think of "distant.") The break on the femoral fragment was fresh enough for us to hope that the telltale end of the bone might be somewhere nearby, perhaps right underneath. But we kept on moving up the gully. An hour later we came across a piece of tibia, or calf bone, the same dark color as the other remains of our hominid stranger—another good reason for high spirits.

The tibia and femur fragments both had another, extremely intriguing feature in common with the arm bones we'd discovered already.

"Don, you know Lucy's skeleton better than you know your own face," said Tim when we broke for lunch. "Now take a look at this thing."

Tim unbagged the piece of tibia and held it up. We were sitting at a worktable in the lab building, an airy stone structure outfitted with floor-to-ceiling shelves crammed full of mammal fossils and boxes of stone tools. In the middle of the room, a fiberglass cast of the Laetoli footprints was propped up on sawhorses.

"It's a little bugger, isn't it," I replied.

"Small as Lucy?"

"Yeah."

"Maybe *smaller*?"

"That doesn't seem possible. A full-grown hominid running around less than three and a half feet tall?"

"Who knows? But it would be nice of Lucy were here to compare it with."

"Kimbel is due here from the Institute in a couple of weeks," Gerry Eck reminded us. He was stooped over a wash pail on a table nearby, scrubbing the dirt off the morning's gleanings from the site. "If we can reach him in time, Bill could bring casts of Lucy's limb bones with him."

Over lunch—potted meat, cabbage soup, and Kool-Aid—we

drafted a telegram to Kimbel, to be sent out with the first vehicle that came through.

HOMINID IN THE BAG, we wrote. BRING LUCY'S RIGHT ULNA, CERVICAL AND LUMBAR VERTEBRAE, PROXIMAL FEMUR, PROX AND DISTAL TIBIA, PROXIMAL LEFT HUMERUS. Then we detailed some needed supplies that he should bring with him too. Heading the list was TWO POUNDS OF PEANUT BUTTER.

The sun was still high, so there was no point in returning to the site for another hour. Tim gathered his students and we went back to the lab to begin sorting the surface material. He picked up a bag marked "Gully One" and dumped the contents out on the worktable.

"Gerry's washed this stuff, so it's a lot easier to see what we've got," said Tim, picking some pebbles out of the pile and flicking them out the open window. "The next step is to go through each bag piece by piece and isolate anything that has a remote chance of being hominid. If we make a mistake and put a piece of the hominid into the nonhominid pile, we'll never retrieve it—that piece might as well have been washed away to the ocean. Sorting is too critical a process to leave to any one person. Before I relegate anything to the nonhominid pile, I'm going to hand the bone to Johanson. If he has any doubt, then we keep the bone in consideration."

Tim picked out of the bag a concave shard of bone that to an untrained eye might look like a piece of cranial vault.

"I say this one's out," said Tim, handing me the specimen.

"Right. Turtle shell," I said.

"What about all this pale bone?" asked Berhane.

"We can't judge by color yet," said Tim. "Occasionally a fossil has multicolored parts."

One by one we went through the bags brought back from the site. We didn't talk much. We checked the bones for texture and quality of preservation, as well as for hominidlike morphological features. When we came across bones that belonged unquestionably to some other species, we placed them in a separate bag. Information about the fauna commingling with the hominid would help us get a fix on the ecological conditions prevailing at the site. Tim held up each piece to the light of the window, turned it in his fingers, and pronounced his decision: *In . . . Out . . . In . . . Out . . .* In his deep, sepulchral voice it sounded as if he was ruling on the fate of souls. He handed each bone to me, and when there was the slightest shadow of doubt I questioned his verdict. By 2:30 we had gone through most of the bags and had accumulated perhaps a hundred bones that *might* belong to our hominid. These would be part of our raw material when we started trying to put the skeleton together. That would be a far more challenging process—but first we had to retrieve whatever further information the site would yield. With luck, it would overwhelm what we already had in hand.

If the morning's surface search had been promising, the afternoon's was unqualified fulfillment. Shortly after we returned to our positions on the line, Tim plucked a fine piece of palate out of the gravel by his knee. Later, back in the lab, it would fit neatly onto the maxilla found the night before. More finds followed, each one strengthening our hopes. Berhane uncovered the first scrap of lower jaw. It included an eroded remnant of a bump of bone called the mandibular condyle, where the jaw hinges with the skull. Meanwhile fragments of tooth roots seemed to be everywhere. Then Gerry turned up another stretch of limb shaft.

"The rest of the ulna!" Tim exclaimed. "Way to go Eck!"

"We better rewrite that telegram to Kimbel," I said. "Tell him to bring Lucy's whole skeleton."

Berhane's luck continued: Late in the afternoon, he held up what appeared to be the first piece of tooth crown. It wasn't enough to give an indication of the species we were dealing with, but it was a promising way to end the day. We finished the surface search with a little light to spare. As the shadows gathered, we set up for the next day, using twine and nails to divide the entire site into a grid of two-meter squares. Tomorrow we would begin the next step: sweeping the loose lag out of the squares for sifting, one by one.

At five-thirty the Land-Rover arrived from camp. It would require two trips to take us all back, and I volunteered to go in the second load. While I waited, I climbed up the steep slope just to the west and sat down on a comfortable ledge. Looking down at the site, I was suddenly struck by its overwhelming obscurity. Against a background of magnificent cliff faces flushed by the sun lay this stretch of gray nothingness, littered here and there by a spindly wisp of a bush and scarred by the road running along its flank. For decades now, that road had carried tourist combis around the corner to view the plaque erected at FLK, marking the discovery of *Zinjanthropus*. As the combis passed, passengers on the right side would enjoy a view of the cliffs and the river margin below, with its brilliant patches of yellow-green sisal, graceful acacia silhouettes, and stands of aloe vera, their leafy heads perched on rough-barked stalks ten feet tall. Passengers on the left, meanwhile, would have a view of what we were now calling Dik-dik Hill: a brief, apologetic rise in the bare terrain, with a mop of vegetation slapped off center on the summit like a cheap, ill-fitting toupee.

On one side of the combis, in other words, lay Africa, while on the other, the scene resembled a vacant lot. I couldn't help smiling at the irony. The tourists had come to see Africa, of course; but it was the vacant lot that held the treasure—this enigmatic ancestor, all those years lying right out there on the surface. And now we had found it and dressed up the hill with a geometric grid, each square brimming with possibility. Specimens that clearly included both skull and limb bones of either robust australopithecines or *Homo habilis* were not merely rare in the fossil record—they were nonexistent. Thus there was an enormous gap in our understanding of the basic anatomy of these two species, not to mention the problem of some very fuzzy attributions of limb bones and other postcranial parts. Were the isolated femurs found at Koobi Fora really those of *Australopithecus boisei,* or was that just a guess?

And what about *habilis*? OH7, the type specimen, consisted of some skull fragments, part of a mandible, and a handful of hand bones. Were these really part of a single juvenile *habilis,* as the Leakeys believed? What about the foot bones found at the same site: Did they belong to the same individual, or to an adult female? On the Zinj floor itself, the Leakeys had turned up some pieces of tibia and fibula. Did these belong with the other *habilis* postcrania, or were they pieces of Zinj himself, the original *Australopithecus boisei*? Which bones went with which species? It was all too confusing to yield much insight.

If the Dik-dik Hill hominid did no more than help us pin down the identity of such specimens, we'd still be way ahead. But getting the names straight on fossils is always just the beginning, the means to approach deeper questions. Were the robust australopithecines really "robust"? Was *Homo habilis* proportioned like us, as most people thought? Did it walk like *Homo sapiens*? What was the relationship between brain and body size during this critical transitional period in human evolution? Some answers might already be resting on museum shelves in Nairobi, Johannesburg, or Addis Ababa, needing only the right key to unlock their ambiguities. If we could find more of this specimen, it could serve as a sort of Rosetta stone, making sense out of a jumble of uncertainties. Who knows, I thought—maybe Dik-dik Hill would have its own plaque someday.

Back at camp I found Tim and his students at work in the lab, hunched over the fossils like a trio of alchemists about to conjure up the philosopher's stone. They hadn't even bothered to wash up first. Tim looked up and grinned.

"Nothing in the world I'd rather be doing than putting a robust australopithecine skeleton together," he said. "You're just in time to watch these ulna pieces match up."

Tim held one piece of elbow bone in each hand and gently clicked them together. The point of contact was narrow, but there was no doubt about the fit.

"So you're certain it's a robust?"

"These are big teeth, bwana."

"We still can't really be sure of that," Gen interjected.

"Mr. Suwa here has been arguing for *habilis*. He hasn't learned yet that a grad student is supposed to agree with everything his teacher says," Tim said facetiously. "Berhane is a lot smarter. He thinks like me. This hominid's a little robust. Probably female."

I picked up the portion of mandible and ran my finger over the slight, eroded protruberance at the back.

"What about this little condyle?" I said. "It's tough to imagine a great big robust jaw full of megamolars swinging from a joint that size."

"So you think it's *habilis* too?"

"Let's just say I suspend judgment."

"OK, we've got two votes for *boisei*, one vote for *habilis*, and Johanson here is being wishy-washy. What all this tells me is that we need to find a damn premolar crown, bad. Or a molar crown."

"While you're taking orders, how about the missing part of this femur," I said holding the lower limb fragment up to the light. "This break is real fresh."

"We'll get it all. Tomorrow."

We were back on the site early on July 23, eager to begin the sifting operation. I've mentioned before what an advantage it is to have a good, interdisciplinary team in the field. In this expedition the core disciplines—paleoanthropology, paleontology, archeology, and geology—were superbly represented, in some cases by people who had to be considered among the best in their fields. Our immediate need, however, was not heavy credentials, but an ability and willingness to haul dirt. Our funding did not provide for paid laborers. Everyone pitched in with spirit, including camp visitors like Jeremy Paul and science writer James Shreeve, who was staying with us for a few weeks. Tim and Gerry worked the brooms, stooping over to sweep the loose dust and gravel on the hillside into the shallow karais. The rest of us took turns carrying the laden pans to a level area near the road, to be dumped into the sieve. Two others lifted the wooden tray and rocked it back and forth to coax the dust through the screen. If the sievers were lucky, the wind would carry the dust away, but whenever the wind stopped, the dust puffed up and enclosed them in a blinding cloud.

Once the dust was fully expelled, the sievers dumped the bone, rock, and other material trapped in the screen into another karai. The remainder of the team squatted on the ground, tossing the sieved material one handful at a time into karais, where it could be shaken, spread out, and scanned through, pebble by pebble. Every splinter of bone was picked out and put into a wooden box. It was not work that required a Ph. D.—just a good pair of eyes and a lot of patience. By lunchtime we had completed three squares and found nothing that could unequivocably be attributed to the hominid.

The only scientist excused from sifting work was Bob Walter. The geologist had his own labors to attend to. After spending the first day getting his bearings in the general framework of the site, Bob now had the difficult task of determining as closely as possible where in the Olduvai stratigraphy these hominid bones were emerging from. The gravel forming the surface of Dik-dik Hill was a colluvial lag—a term for the accumulation of debris at the bottom of a slope. This meant that even if we were to find some fossils embedded in the lag, rather than lying on top, they could still not be considered *in situ,* since the lag itself was not a primary deposit, but rather a composite formed of material from any number of layers upslope. With Bob's help, however, we might be able to locate the primary deposit nearby, where the hominid had originally come from.

The day before, Bob had quickly identified two key marker tuffs above the site in the surrounding terrain—Tuffs IC and ID. Though much the same color, the two tuffs could be easily distinguished if you knew what you were looking for. Tuff ID was characterized by horizontal layers of pumice chunks, dotting the ash like raisins in a cake. Tuff IC was finer-grained and showed no such horizontal layering. Toward the end of the day, Bob also got a fix on Tuff IF, a key ash layer that marks the boundary between Bed I and Bed II. Of the three, Tuff IF was the only one that had been dated using the potassium-argon method. The experiments had come up with some rather shaky answers: Tuff IF was anywhere from 1.75 million to 8.5 million years old. Obviously the latter date was incorrect.

"Olduvai is such a classic site that people forget it isn't all that securely dated," Bob reminded me recently. "Most of the Olduvai tuffs aren't very pure—they're contaminated by older rock that probably erupted from volcanoes along with the primary ash. The only firm potassium-argon date is still the 1.8-million-year mark that Garniss Curtis and Jack Evernden got for Tuff IB, in the experiments in the late fifties that broke open the field. Above Tuff IB, you've got a million and a half years of stratigraphy without any firm age control."

Getting some control on the ages of the Olduvai horizons was one of our long-term goals, and I was counting on Bob's expertise. He would have the substantial help of some new radiometric techniques being developed by Garniss Curtis and Bob Drake back in the Institute's dating lab. In the meantime, the Olduvai hominids weren't swimming in a completely uncharted ocean of time. Several other dating techniques had been tested over the years at the Gorge and had lent support to some otherwise rickety pottasium-argon dates.

The most informative of these relies on a fascinating property of the earth itself, called paleomagnetism. For reasons no one really knows, the globe acts like a gargantuan magnet. Like other magnets, the globe possesses polarity—it has a negative North Pole and a positive South Pole. To anyone who has watched iron filings arrange themselves under the influence of a bar magnet, it should come as no surprise that as molten volcanic rocks cool, crystals in the rocks also align themselves with the earth's magnetic field. As the rocks solidify, the crystals retain their orientation, like compass needles that have been locked in place. Thus every magnetic mineral in every rock all over the world should point to the North Pole.

In 1906, a French physicist named Bernard Brunhes made the perplexing discovery that *some* volcanic rocks were magnetized in precisely the opposite direction—toward the South Pole. Brunhes concluded that the earth's magnetic field in his study area must have somehow reversed itself. The concept of reversing magnetic fields did not attract much attention until the early 1960s, when three geologists, Alan Cox, Brent Dalrymple, and Richard Doell, became intrigued and began to sample the magnetic polarity of basalt samples from scattered places around the earth. Along with other scientists, they eventually proved that no fewer than nine flip-flops in the earth's field had taken place over the past three and a half million years. And whereas Brunhes had believed that the curious reversals were a local phenomenon, the studies in the sixties showed that the effect was consistent the world over. No matter where the scientists took their samples, the indelible magnetic record of the earth fell neatly into four "polarity epochs," punctuated by much shorter periods of reversed polarity, which they called "events."

The first one of these quick-shot reversals to come to light is called the Olduvai Event, named after a busy little canyon cut on the edge of the Serengeti Plain. Discovered by Hay and Grommé in 1963, the Olduvai Event is a period of normal, northward polarity in an upside-down time called the Matuyama Epoch. Hay and Grommé found that the reversal in polarity lasted through the whole of Bed I times, on up into the lower part of Bed II. There is nothing intrinsically "datable" about a particular paleomagnetic flip-flop: All you know is that it happened before some flips, and after other flops. But since the *order* of events is consistent, if you can pin an absolute age on an event at one location by some other means, then you've got a date for that event that applies the world over. In 1972 geologist Neil Opdyke analyzed the magnetic orientation of some deep-sea cores from the ocean bed at several different locations around the world. He found the Olduvai Event clearly and consistently recorded. By comparing the paleomagnetic data with known rates of sedimentation of the sea bottom, Opdyke figured out that the Olduvai Event lasted from approximately 1.85 to roughly 1.7 million years ago, a period of 150,000 years.

That evidence from paleomagnetism fit very nicely with the potassium-argon date of 1.8 million for Olduvai Tuff IB, which is found near the bottom of the Olduvai Event at the Gorge itself. It also gave some muscle to estimates for the rest of Bed I—including Tuff IF at the top, and IC and ID sandwiched between. According to this technique, at least, all of the tuffs under Bob Walter's scrutiny were at least 1.75 million years old, and no more than 100,000 years older than that.

At the hominid site, Bob Walter's next step would be to try to

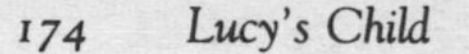

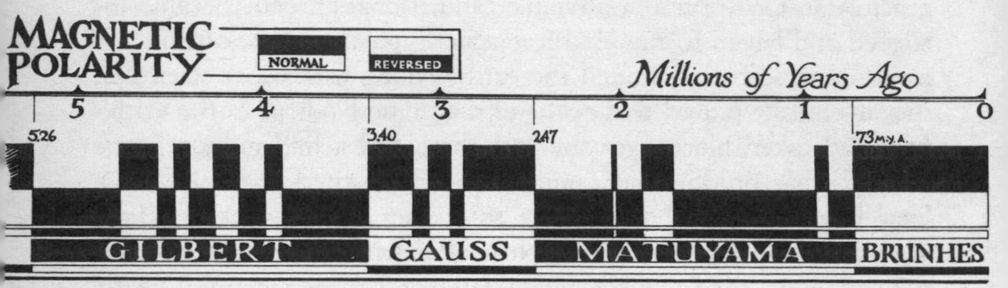

Paleomagnetism is another technique used to provide a chronological framework for ancient deposits. Crystals in cooling molten rock line up with the earth's magnetic field, like compass needles. Over the millennia, the magnetic field of the earth has reversed itself periodically, so that rock crystals point either to the North Pole (normal) or to the South Pole (reversed). The four major polarity chrons—the Gilbert, Gauss, Matuyama, and Brunhes—were punctuated by shorter-term reversals called "events." The Olduvai Event in the Matuyama Chron lasted from 1.85 to 1.7 million years ago.

use these markers to place Dik-dik Hill itself in the larger context of the Olduvai stratigraphy. This would require digging a trench into the steep slope adjacent to the site, exposing the series of deposits that might once have lain over Dik-dik Hill but had subsequently eroded away—leaving behind only the gray shroud of colluvium that we had so suddenly taken an interest in. Bob would then have to dig another trench into Dik-dik Hill itself, hoping to find some correlations there with the deposits exposed by his first excavation.

Most geologists will tell you that digging trenches is not one of the relished delights of the profession. It helps to have some company around—preferably somebody with a strong back. Unfortunately, Bob's student Paul Manega had received notice the day before that he was wanted back in Arusha, and had hitched a ride out of the Gorge with the departing lorry. We could spare no one from the sifting operation, which left Bob to dig alone. He didn't seem to mind at all. Every once in a while through the day I would look up from my karai-ful of pebbles and dirt and I'd see Bob, stripped to his Patagonia khaki shorts and a layer of sunscreen, swinging his pickax into the slope with all the gleeful gusto of one of the Seven Dwarfs. Sometimes I could actually hear him whistling while he worked.

In the afternoon some more hominid tooth-root fragments began to appear in the karais, but not much else that we could definitely put a label on. The day before, the plethora of root fragments had seemed encouraging. Now it was beginning to worry me. One reason that teeth play so large a role in distinguishing between extinct species is their relative abundance in the fossil record. Tooth enamel is harder than bone, and thus is more likely to withstand the pressures of time. But *this* skeleton's dentition had exploded all over the place—which did not augur particularly well for the preservation of softer parts. I'd feel a lot better, I thought to myself, if we could find one intact tooth. Or anything else that would tell us what we were dealing with.

By four o'clock we had swept clean the surface of several more squares on the slope of the hill just below the dik-dik pile. We were dust-caked and bone tired, but had uncovered a curiously meager collection of fossil fragments. Only a dozen or so of the bone chips we'd found were big enough to warrant attention, and of these, none at all belonged unquestionably to the hominid. After yesterday's success, the disappointment was all the sharper for being wholly inexplicable. Tim glared at the cleared square at his feet, looking somber and puzzled, as if he held the ground personally responsible for thwarting his expectations.

"It's late," somebody said. "Let's call it quits."

"There's still light enough for one more square," Tim insisted. "I got a good feeling about the next one up the slope."

He stooped over and began to beat the ground again with his broom. We set to work with the sieve and karais. By the time Bob Walter came by half an hour later and plopped himself down near the pile of dik-dik pellets on the top of the knoll, we were three quarters through. But by then it was clear that this square too was strangely infertile.

"I haven't heard a lot of noise from you guys this afternoon," Bob said, taking a drink from his canteen.

"Whatever the hell this hominid is, it's not being very cooperative," Gerry Eck said.

"Tomorrow we'll finish sweeping the squares down the gully," Tim muttered, still working his broom. "If we get something, I think we should start excavating down in the lag." Then he added, almost to himself, "If nothing turns up tomorrow, I wonder how much time we want to put in here."

If that was a question, nobody bothered answering it. We finished up the sifting and slowly gathered the equipment together. Bob Walter sat perched on the knoll, toying with the strap of his canteen and staring blankly into the dik-dik patch. Even he seemed beat by the day's efforts. But then I noticed Bob's eyebrows lift up a bit, and his gaze gather interest and surprise. He leaned down and peered into the pile of dried-up feces, as if they held some message of deep significance. He bent over and plucked something out of the dik-dik latrine.

"Holy *shit*," said Bob. In his hand was a hominid canine tooth, perfectly intact.

Simple gratitude demands that I pay some passing respects at this point to the dik-dik. These pretty little antelopes are a frequent sight in the Gorge, and would be even more noticeable if their brown-and-gray-flecked fur didn't match the stratigraphy so well. If you should visit the Gorge and catch a glimpse of one scampering up a steep slope, quickly look for another. Dik-diks mate for life and stick close by their partners. The females of the species are larger than the males. Both sexes have big, fleshy noses overhanging their mouths, giving their faces an expression of perpetual goofiness. (The oversized proboscis acts as a radiator to cool the blood.) Dik-diks live on highly nutritive but widely scattered foods like buds, fruit, and young green leaves. Each monogamous pair must therefore reserve for itself a large and well-defined territory. They remind neighboring dik-diks of their claim with insistent scent markings. The males are especially well equipped for the task, their faces adorned by quarter-sized scent glands just below the eyes, which they use to daub a scenting fluid onto particular plant stems in the territory. Diligent maintenance of these territorial signposts envelops the plant stem in a sticky ooze.

Another territorial trick of the dik-dik is the persistent use of specific defecation spots, anthropomorphically referred to as latrines. Typically, the female will use the facility first. When she is through the male will paw and muzzle her leavings before unloading his own. The ritual, like much of observed dik-dik etiquette, probably reinforces the bond between the pair. Rarely—very rarely—dik-dik feces will also mark the location of priceless hominid fossils. Whether such warm attention given to ancient bones represents an integral part of dik-dik behavior or is merely coincidence is too difficult to ascertain from the data currently available.

Bob's discovery was enough to send us back to the lab that evening with rejuvenated spirits. But while the tooth Bob had pulled

Bob Walter poses behind the pile of dik-dik droppings from which he pulled a 1.8-million-year-old hominid canine tooth.
DONALD JOHANSON

from the pile was a breakthrough, it was also a source of further frustration. The canine and its root were preserved exactly as they had been when the individual died almost two million years before. Missing from the tooth, however, was what had been lost in life. Almost the whole of the crown had been worn down through years of chewing. Clearly, the hominid had been pretty well along in life when it perished. Judging from the root and what remained of the crown, we could also tell that this canine was surprisingly large. This was an important clue in our search for the hominid's identity. Remember, while robust australopithecines have big molars and small front teeth, in early *Homo* the situation is just the reverse: It tends to have relatively *large* canines and incisors, and relatively small cheek teeth. To make a certain diagnosis, however, it would have helped to know the height and shape of the canine crown. That information was lost forever.

"I'm sorry, Bob, but we're going to have to send you back into the shit to find another tooth," Tim laughed. "Bring me back a big, beautiful back molar. Otherwise I might have to buy the crock that Suwa is dishing out and start believing we're chasing a little *habilis.*"

"With a canine root that size its *got* to be *Homo,*" Gen said. The three of us were clustered around the bones with only a couple of flashlights illuminating the work. Berhane was feeling ill and had retired early.

"Not necessarily," Tim said. "Look at the bicanine distance. It's really reduced."

The day before we had put together enough of the upper jaw to be able to estimate the distance between the two canines by the position of their root sockets. In this specimen, there wasn't much space between the canines to fit in the large incisor teeth characteristic of *Homo.* It was just another hunch, of course, but the small bicanine distance would lend support to the notion that our hominid was a robust australopithecine after all.

"Yeah, but look at the back edge of that incisor socket," Gen countered. Having spent much of the last six years becoming intimate with the arcanities of hominid teeth, he wasn't about to relinquish the high ground. "That's bigger than any I've seen on a robust max."

"OK, Suwa, forget the teeth for a minute and check out the guttering of this nasal margin," Tim challenged him, tracing with a finger the contour of what little remained of the side of the nasal opening. "Wouldn't you say it flows smoothly into the nasal cavity?"

A smooth, rounded edge on the nasal margin was another distinct characteristic of a robust. Gen took the bone and studied it from different angles. He wasn't as familiar with nasal morphology as he was with teeth, and I could see that the question bothered him. He pushed his lips forward in a pout of concentration, struggling to get a fix on the anatomy. I noticed that Tim was looking at Gen too, studying his face. Suddenly I realized that Tim was baiting his student, pushing on Gen's memory and conviction to see how well he could defend his position.

"You can't really say that," Gen finally announced. "It could be that the margin's simply eroded. We'll have to find more bone before we can make a judgment."

"Spoken like a born paleoanthropologist!" Tim laughed and slapped his hand on the table, relieving the tension. "When in doubt, shake your head and call for more fossils. OK, we'll say it's *habilis* tonight. But tomorrow I'm gonna find a molar crown under that dik-dik pile big enough to land a helicopter on. Now let's get some rest."

In our map of the site, the two-meter area where the canine had been found was labeled Square 8. The next day we resisted the temptation to dive right into it, and instead spent the morning processing the remaining loose lag. Berhane, stricken by some sort of stomach flu, was on his back in his hut, and without him the work went that much more slowly. Five hours of sifting produced one mottled ridge of palate bone and little else. In the afternoon we cleared away the dik-dik pellets and the other surface detritus from the magical eighth square. Gingerly at first, Tim began to poke into the compacted lag with his rock hammer, trying not to damage any fossils that might be lying beneath. The colluvium proved thicker than we had thought, which meant more dirt for the karais. Nothing was turning up. By three o'clock it was clear that the square was not so chock full of fossils as we'd hoped. Tim muttered and swore and picked up his pace. An hour later we had to concede the truth: The square was barren. The dik-dik had let us down.

That evening in the lab we discovered that the piece of palate we'd found in the morning fit neatly on to the back of the maxilla. It was a small, sustaining consolation to the afternoon's disappointment. And that was to be the pattern over the next several days—we'd bring hopes to the field every morning, and gradually they would dissipate and blow away like dust from the sieve. Then, just when we began to think that it would be prudent to cut our losses and abandon the site, a teaser would turn up in somebody's karai: a bit of cranial vault, a good root fragment, another little stretch of limb shaft. Nothing that would reveal the identity of the hominid, but enough to give us legitimate hope that the next day would yield up that elusive diagnostic piece.

In the lab one day I found Tim staring at the hominid's humerus. A beautiful inch-long chunk of shaft had turned up in somebody's karai that morning, and Tim had glued it neatly in place.

"That's a nice join," I said. Tim didn't look up. He just stared at the bone. For a moment I thought he might be onto something. But he just sighed and leaned back in his chair. Suwa came in and starting poking through the morning's yield of fragments. He handed Tim a thick splinter of limb bone.

"What do you think?" said Gen. "Maybe tibia?"

Tim glanced at the fossil and dismissed it with an impatient wave of his hand. Then he leaned forward again and practically glowered at the humerus. I'd seen Tim in these dark moods before, and knew better than to intrude. But Gen's natural enthusiasm wouldn't allow such circumspection. He wanted more feedback.

"I'm not saying this bone is a hominid tibia," he said. "Just that it might be."

Tim grabbed the fossil from him and pointed to a groove running along its length.

"What do you call this?" he said. "Ever seen a hominid with a groove like that?"

"I admit it's a little weird."

"Weird is right. You've got hominid blindness."

"I didn't say it *was* hominid . . ." Gen insisted.

"And I'm not saying it isn't," Tim snapped. "Just that it's a real low probability. So save it for later." Then, perhaps sensing Gen's pained surprise, he softened his tone.

"Forget that piece of junk, Gen," he said. "Go out and find the proximal end of this humerus. Maybe that'll help us understand the *length* of this damn bone. Because from what I can see now, it isn't making any sense."

Over the next couple of days we expanded the quadrangle of squares at the site and began to dig on the other side of the shallow gully. When that proved fruitless, we excavated on the north flank of Dik-dik Hill farther down toward the road. Those squares were a little richer, confirming our belief that the hominid's bones were eroding out of the hill itself, rather than tumbling down from the steep Gorge slope overlooking the site. Meanwhile Bob Walter had dug a trench into the barren south flank of the hill. He found that the rock fragments in the fossil-fertile lag layer were identical to those found in Tuff ID and below. The absence of any material in the lag originating from higher levels supported the belief that the hominid bones too were no younger than Tuff ID itself. In the absence of an *in situ* find, it was again only an inference. But it was a strong one.

Bob and Paul Manega had done about as much as they could for the time being. They returned to general geological surveying on foot in the eastern part of the Gorge, keeping sharp eyes out all the time for a small pride of lions that they had run into earlier near the Second Fault. Any lion who had stayed in the Gorge over the dry season, instead of following the migrating herds of wildebeest across the Serengeti, was bound to be hungry and unpredictable.

For those of use working Dik-dik Hill, the worst threat was the numbing tedium. After a week, the work had fallen into a rhythmic redundancy—spill your handful of pebbles into the karai, shake it, scan it, spill it out, grab another handful. Other than our own conversation, the only distraction was the occasional tourist combi rolling by on its way to FLK. Every once in a while, the overlook at the museum half a mile away would fill up with bright colors—the T-shirts of the tourists stopping for the view—and we would hear their excited voices filtering across the Junction. Twenty minutes later their vehicle would come rattling down into the Gorge, suddenly appearing around the bend and heading right toward us. The combis slowed down as they passed, then disappeared in the fog of gray dust kicked up by their wheels. I had given instructions to Peter, the guide at the museum, to keep the public away from the work area. That was simply good practice. But now I found myself secretly hoping that the combis would stop to allow their passengers a chance to ask us some questions. Or I could ask them some. Anything to relieve the boredom.

Tim's mood was changing too. His initial sense of triumph at finding the first fragments so quickly ("The Leakeys took thirty years to find a hominid at Olduvai Gorge," he told a visitor. "We got one in three days") had been replaced by a clenching determination to dig as much information as possible out of the ground before the field season was over. Hour after hour he stooped over, legs spread apart, shirt drenched with sweat, scratching at the earth with his rock hammer, then shoveling the loosened dirt into a karai. If the hominid's identify would not spring forth voluntarily, then Tim seemed ready to squeeze it out of the ground by force, one fragment at a time. He worked with such intensity that sometimes even six or seven people sifting through the material were not enough to keep up with him, and the panfuls of unprocessed material would stack up by the sieve. Rather than use the opportunity to take a break, Tim would drop his hammer and squat down to help us, casting handfuls of dirt into a karai and scanning through them with a silent, machinelike precision, then dumping the meaningless rubble out in disgust.

One afternoon the screen in the sieve developed a hole—small to begin with, but threatening to rip wider and bring the whole operation to a halt. If any fragment of the hominid passed through, it would be lost forever in the pile of dirt growing below the sieve.

"Have we got any backups?" I asked Gerry. We were standing around the broken sieve, looking as hopeless as a bunch of stock traders whose computer has just gone down.

"Hell, no," said Gerry. "Somehow I didn't foresee the possibility that we'd be sifting through a few tons of gravel this soon. I'll try to improvise a patch. Otherwise we'll have to go all the way back to Arusha for new screen."

To wait for screen from Arusha would mean the loss of at least two days of work—probably more, given the shaky condition of our solitary Land-Rover. Tim turned around and strode back over to the square under excavation.

"Damn hominid broke our only screen," he swore. "Let's get him!"

Tim spread his feet apart again and set to work, putting behind each controlled swing of his hammer a riveting concentration, as if strength of will alone would ferret out the hominid and banish all disappointment forever. I knew that Tim's frustration was not really directed at the fossil, nor at the delay threatened by the broken sieve, nor even at the shortsightedness of granting agencies that had left us out here without adequate equipment. The object of his anger was stronger, darker, and maddeningly elusive—and approachable only through such bald physical intensity. Watching him probe into the earth with his fingernails, it struck me that at this moment Tim's enemy was manifest in the ground itself. This stretch of ancient earth had entered into a pact with us, and was now betraying that pledge. Several days before, in near darkness, Tim's acumen had recognized a little plug of bone for what it was, triggering the metamorphosis of this anonymous, shit-topped knoll into a hominid site of enormous potential. In return, the perfectionist in him demanded of the earth that it yield under his direction and deliver up its promised revelations.

Instead, Dik-dik Hill—Olduvai ground, don't forget, the Leakeys' "backyard"—had chosen not to reveal but to jealously obscure and obstruct. In so doing, it had allied itself with ignorance. For myself, ignorance defines an absence, a lack of understanding. I'm sure the same is true for Tim's other colleagues—even for Lew Binford, for whom the word carries very special connotations. But watching Tim scrape and swear and drip sweat into the dust under his hammer, I realized that for him ignorance was an active antagonist, never absent for very long. It could take shape as intractable earth, or it might peek out from between some grant-review sheets full of damningly faint praise, or it might put on a suit and tie and lecture to the world on the antiquity of man, all the time restricting access to knowledge that might prove otherwise. In any of its guises, ignorance was the enemy, and if you did not fight it, you were part of its strength.

We worked till the shadows dispersed, tilting the sieve away from the hole and rocking it more gently than usual. The day's efforts yielded two small fragments of tooth crown. That evening Gerry patched the screen, and the next morning we found another molar fragment and some bits of palate. The day after, it was a shard of femur shaft, a chip of cranium. The days of hauling and sifting rub-

Tim White at Olduvai Gorge JAMES SHREEVE

ble turned to weeks. Still there was no breakthrough, but slowly, very slowly, the hominid was coming up for us. I must admit that sometimes, while I pushed the pebbles through my karai or relieved Tim on the hammer, I thought wistfully about the expeditions in Ethiopia, where Lucy's bones and the other *afarensis* specimens had lain out on the slopes for us, to be picked up heavy and whole, like ripe fruit.

Excavating this fossil was more like digging seeds out of winter earth. If skillfully handled, the nameless fragments accumulating in the lab might eventually sprout meaning and matter. At this point, it seemed just as likely that their significance would remain forever unexpressed. Every day Tim and I talked about closing down the site and moving to Laetoli, just as soon as we had a vehicle. We both knew, however, that we weren't going to make it to Laetoli, not this summer, not while there was still hope that this creature who so long ago had died on that spot by the Olduvai lake could tell us something new about the way it had lived. We were seduced by the fossil, and would stay at the site for as long as it took. Too many questions were left to be answered.

CHAPTER SEVEN: *The Primitive Wonder*

False facts are highly injurious to the progress of science, for they often endure long; but false views, if supported by some evidence, do little harm, for everyone takes a salutory pleasure in proving their falseness; and when this is done, one path towards error is closed and the road to truth is often at the same time opened.

—Charles Darwin, *The Descent of Man*

BY THE END OF JULY, Dik-dik Hill had lost all of its charm. After the initial excitement, Lew Binford and George Frison had returned to doing what they came for—scouting the long-term archeological possibilities around the Gorge. Eck and Ndessokia were itching to get back to paleontological surveying, and with the returns diminishing at the hominid site, we had no choice but to let them go. The excavation crew was reduced to Tim and his two students, along with Mzee, our lone African staffer, and myself.

I could no longer imagine commemorative plaques being erected on the spot, except to pay tribute to the torments of boredom. My whole past and future seemed reduced to squatting in the dust, picking dark flecks out of an ever-replenishing mound of dirt. I began to take an interest in matters of profound inconsequence—the breadth of silence between each note of a bird's call, the pattern of sweat on somebody's shirt. I grew intimate with each of my co-workers' style of sorting through their portions of sieved detritus. Mzee Mrisho, for instance, sat with his legs spread, hour after hour, his karai cushioned between his thighs. He rooted through the contents of his pan with a forefinger, touching the pebbles one by one. When he came across a piece of bone he picked it out and placed it in the wooden box with an air of proud solemnity, as if he had were returning a stolen gem to its rightful owner.

Gen Suwa's approach was equally formal and precise. He put two handfuls of gravel into his karai, and patted them down with his palms into a neat layer one pebble thick lining the sides of the bowl. Then he raised the whole karai up to his face, studying the dust-covered particles as if each were a character in some ancient text. Berhane had developed a more whimsical technique. He held a handful of gravel up high and let the pebbles trickle out in a stream through his fingers, ringing the karai like a wind chime. When he found a chip of bone, he tossed it into the wooden box with a elegant flick of his wrist. He never missed.

In the mornings, we could usually look forward to a couple of tourist combis rolling by on their way to FLK. By this point even Tim looked forward to the distraction, especially if the tourists took a genuine interest in what we were up to. Unfortunately, many of them seemed reluctant to acknowledge our greetings, much less engage in conversation. I finally figured out that their diffidence rose from our being *outside the vehicle,* which from their perspective defined us as a kind of wildlife—suspiciously communicative, but wild nevertheless. To be fair, the dust blowing out from the sieve had painted us the same dull gray as the ground itself, so perhaps we did not appear quite fully human. But it was a strange feeling, to be so thoroughly objectified. As a combi slowed to a stop, three or four heads wearing safari hats would pop out of the open sunroof, cameras clicking away madly, just as if they had encountered a pride of lions lolling beside the road. Bob Walter suggested we put up a sign: DO NOT FEED THE ANTHROPOLOGISTS.

At the end of one hot, bone-poor afternoon I was standing next to Berhane beside a half-excavated square. Tim had taken a rare break with his hammer, and for a few minutes we had no dirt to sift through. Something in the square caught Berhane's eye. He bent down and picked out an inch-long chunk of limb bone—no doubt a bovid's, judging from the thickness of the bone cortex. This was not a fossil of much importance, but certainly worth collecting. Instead of putting it into the box along with the other fragments from that square, however, Berhane kneeled down and tucked it back into the loosened earth at his feet. He looked up and saw that I was watching him.

"For sieving excitement," he explained with a wry grin.

I couldn't fault him. There was no harm, really, in finding the same bone twice.

A few minutes later we heard the sound of vehicles approaching around the bend. It was unusual for tourists to be in the Gorge this late in the afternoon. The two brand-new A&K Land-Rovers barreling down on us were not behaving like standard safari vehicles either. With horns blaring, they lurched back and forth, then skidded to a full stop right in front of us. Out of the lead vehicle stepped David Koch, one of the board members of the Institute of Human Origins who had helped bring us to Olduvai. He had told me that he was planning a safari to Africa that summer and would stop at the Gorge.

David was dressed in natty khakis and was obviously enjoying himself. The other doors opened and the members of Koch's private safari tumbled out, stretching their limbs after the long ride from Manyara. David introduced his party, which included some close friends and their companions from New York City, as well as a pair of newlyweds from Los Angeles. They all looked fresh, buoyant, and untroubled, like children on a field trip.

Then I introduced my colleagues—glum, unwashed, and staring at the visitors as if they'd arrived from some other planet. Berhane's jeans were sagging down and he had his sweater tied around his head like a fat, artless turban; Tim was glowering, out of habit, from under the rattlesnake band of his battered felt hat; and Gen, bare-chested and scatter-haired, looked like some sort of wild primitive we'd brought in from the bush for physical labor. Only Mzee looked vaguely civilized.

"So what are you guys doing hanging out here by the road?" David asked when I'd finished making the introductions.

"Nothing much," I said. "Just digging up a hominid skeleton."

That broke the ice. Tim put on a grin and started showing people around the site, Gen donned his blue windbreaker, and pretty soon everybody was talking cheerfully and looking like they at least all belonged to the same species. Unknown to us, the support lorry for the Koch safari had arrived earlier, and by the time we got back to camp the front yard was graced with a luxurious little community of tents, shower stalls, and dining furniture. David and his friends were eager to see the hominid, so we went straight to the lab and laid the bones out on the table. Naturally he was ecstatic that his support for the expedition had so quickly borne fruit, and I confess it was a wonderful feeling to have the specimen in hand and a chance to talk about it with new faces, interested people. For the first time since the early excitement at the site, it struck me how *much* we had found, rather than how little, and what a rare stroke of fortune it was that had led us to that fossil. We celebrated with cocktails around a roaring bonfire.

David's visit provided a happy excuse to take a break from digging. With the temporary surplus of vehicles, we decided to drive down to Laetoli. Though we still had no permit to work there, we could at least check on the condition of the footprint site and give the safari a little tour.

When the Leakeys first visited Laetoli half a century ago, the journey from Olduvai took three days. Today the trip takes an hour and a half, even less if you are not too concerned about the longevity of your vehicle. We thumped along, following the Side Gorge as it curved south, steadily climbing up the plateau separating the drainage basins of Olduvai and Lake Eyasi to the south. I sat crammed in the rear of one Land-Rover with Paul Manega and Bob Walter, the view out my window dominated by Lemagrut, flanked on both sides by the great Ngorongoro caldera. These volcanoes are now in their dotage, but in the late Pliocene they were mere molten urgings somewhere deep beneath the earth's crust. The region then was as flat as the Serengeti itself, and with a climate much the same—hot and dry. The only active volcano in the area was Sadiman, now no more than a gentle swelling in the plain, hidden from sight behind Lemagrut. Beginning some four million years ago, Sadiman puffed out a succession of new savanna surfaces, some as fine as filo dough. It was in these famous "Laetoli Beds" that Mary Leakey's team had found the hominid jaws and teeth in the early 1970s. A few years later, they had made an even more sensational discovery: the footprint trails of hominids and other mammals preserved in the hardened ash.

As we bounced along, Tim and I told our visitors about the discovery of the Laetoli footprints and the incredible good fortune that brought them into being. At the end of a dry season some three and a half million years ago, the volcano Sadiman puffed out a series of ash falls over a period of a few days. After a rainfall, these soft surfaces were walked, hopped, and slithered through by all the animals that lived in the area, from elephants to millipedes. Mixed into the fine sandy material that Sadiman was belching over the plain was a very rare ash called carbonatite. When it meets with water, carbonatite goes into solution, and when the water evaporates, a mineral crystal called trona is left behind, which hardens like concrete. The rain that fell was just enough to dampen the ash, so that the footprints were preserved. Then Sadiman erupted again, laying down a protective layer of new ash. The whole process was repeated several times before the deluge of the rainy season began in earnest.

Three and a half million more years of deposits, faulting, and erosion, and the prints were back out in the open, vulnerable once again. Fortunately, they were discovered before they could be destroyed by erosion. Tim was intimately involved in their excavation. It's a touchy subject for him, to say the least.

"The Laetoli footprints are probably the most precious discovery that has ever been made in this science, or ever will be made," he told our visitors. "It took the most unlikely set of circumstances to produce them. We arrived in a tiny opening of time to find them before they were destroyed forever. You can't imagine what luck that was. You can't imagine what a responsibility rested in their preservation. The restoration of the Sistine Chapel—damn it, the *painting* of the Sistine Chapel—was nothing compared to this. And guess what. Mary blew the job."

Tim had joined Mary's team at Laetoli in the summer of 1978, soon after we had completed our analysis of the Hadar and Laetoli hominids and agreed to name *afarensis.* The original discovery of animal tracks had been made two years before, but it was not until 1978 that a verifiably hominid print was found. One day Paul Abell, a geochemist at the site, came across a broken impression that he thought might be such a print. Tim concurred, but anthropologist Louise Robbins, whom Mary had invited to Laetoli specifically to work on the prints, was certain that the impression was actually two antelope tracks superimposed. She convinced Mary that excavating further under the tuff to look for more prints was not worth the effort. After some vehement argument, Tim and some others in camp prevailed upon Mary to at least send Ndibo, one of the African workers, out alone to the site. Ndibo dug for a while and uncovered two beautiful prints. Unmistakably hominid.

At that point Tim took over the excavation. The footprints were extremely fragile, and to uncover them unharmed he had to develop a new technique.

"The footprints had been made in one layer of ash," he explained, as our Land-Rover headed over to the spot where this had all taken place. "But they had made a dent in the thin ash layers underlying them too. The farther down you went from the actual print, however, the more deformed the impression would be. So we first had to be absolutely sure that we did not puncture through that primary surface—or else we'd end up with only a distorted image. Luckily, there was a thin skin of calcium carbonate lying right over the footprint layer, just a little harder than the light gray ash that fell later and filled in the depressions. I poured paint thinner onto the print, and while the gray infilling soaked it up and darkened, the calcium carbonate skin didn't. In that way I was able to discriminate by color between the two ash layers, and carefully remove the infilling without working down into the print itself."

Using this slow, exceedingly painstaking process, by the end of the summer Tim and the other excavators had uncovered a trail of what proved to be the tracks of at least two hominids, and possibly three. Two parallel tracks, labeled G-1 and G-2/3, ran some thirty feet before they were interrupted by a fault line. A small trial trench dug beyond the interruption revealed that the trails continued on the other side, awaiting further excavation the following summer. But Tim would not be in charge any longer.

"Between the 1978 and 1979 field seasons Don and I announced *afarensis*." Tim grinned. "I was finished at Laetoli as long as Mary Leakey was the boss."

With Tim gone, the responsibility for excavating the much-better-preserved footprints in the southern part of the trail was shared by Mary and Ron Clarke, an experienced British paleontologist. Tim had shared with Ron the paint-thinner technique the summer before. Ron continued in the same manner, etching out the impressions with a fine dental probe, working with his eyes practically inside the prints. Meanwhile, Mary chose to use a mallet and a chisellike probe to remove the hardened ash that had filled up the prints. By the end of the season, they had expanded the excavation to include more than seventy recognizable footprints. The length of the trails covered almost eighty feet. Ron made a mold and brought it back to Olduvai for casting.

The hominid footprint trail at Laetoli. Discovered in 1978, the footprints proved conclusively that early hominids had walked upright three and a half million years ago.
Copyright © by PETER JONES

Mary still had to face the peculiar dilemma of how to preserve the fragile evidence that her team had exposed. She considered erecting an open-air museum on the site, but unless it could be permanently staffed, a building would serve only to attract the attention of animals and vandals. Mary's only other choice was to rebury the whole trail again, covering it over with layers of sand, plastic sheeting, and lava boulders to protect it from damage from plant roots. And so this remarkable relic of the past was brought to light, witnessed, and returned once again to the earth.

On a brief trip to Laetoli in 1985, we had discovered that Mary's efforts to protect the prints from root damage appeared to have backfired. The footprint site was actually more overgrown than the surrounding terrain. There were even a number of acacia trees growing right over the buried prints. Apparently the microhabitat of the disturbed soil had been improved, rather than hindered, by the human alterations. Of course, we could not tell if any damage was actually occurring to the footprints beneath, but it seemed at the very least a risk.

As we approached the site now, I was shocked to see how much the trees had grown in just a year. Some were almost eight feet tall.

"Welcome to Laetoli National Forest," Tim said grimly, as we got out of the vehicle. Our visitors took a few pictures, then explored around a little, keeping clear of the bushier areas where puff adders might be hiding. After a few minutes George Bunn, a friend of Koch's, walked up to where Tim and I were standing and asked a question that was probably on everybody's mind.

"Is this, well, all there is?"

"Amazing, isn't it?" Tim replied. "People have written reams about this site and what it says about early hominid locomotion—their stride lengths and gait values, toe impressions, weight transfer, mean velocity, on and on. If you believe what some people say, from what you see around you we can tell when the Laetoli hominids curled their toes, exactly how their walk differed from ours, and which of them was blind, pregnant, homicidal, or all of the above."

"But there isn't anything here to see," George said.

The sharp lines of Tim's face opened up into a wide grin.

"Bull's-eye. For all that's been written about these prints, only a tiny handful of experts actually had a chance to see them before they were shoveled over. I was one. Ron Clarke was another. Most everybody else who talks about their meanings and implications is talking from casts, not originals. And that's a real shame. A cast can't tell you anything about texture, about color, about the fine details of the original. It doesn't tell you the difference between the actual footprint and the distorted image below it. It doesn't tell you whether some swelling in the print is part of the morphology of an ancient foot, or just a bit of unexcavated infilling."

Tim paused. His grin turned hard, almost menacing.

"But you know what the casts do tell you?" he said. "They tell you where somebody has cut through them with a chisel. Those marks don't come off. They're part of the record now too."

In my science, discovery is always an act of destruction. Whenever we remove a bone from the earth or dig into and around a collection of ancient artifacts, we contaminate, however slightly, the very truth about the past that we seek to apprehend. Nobody, not even Mary Leakey herself, would argue that the mallet-and-chisel method she used to excavate the footprint trail in 1979 was as precise as the technique Tim had devised the year before and passed on to Ron Clarke. The question is how much it *mattered*. If Mary's technique could recover the prints in the same condition as Tim's could, then perhaps he was being overmeticulous. If, on the other hand, damage done to the prints during excavation was influencing the way people were interpreting their meaning, then clearly the record needed to be set straight.

One of the outstanding issues concerned the number of individuals who were represented in the two trails. The G-1 trail had clearly been made by a single, relatively small hominid walking along on two legs. The G-2/3 trail was more ambiguous; most researchers, Mary included, believed that it represented the tracks of two more individuals, the second walking so as to leave prints partially overlapping those of the first. But some workers, including Ron Clarke, believed that only one individual had made the G-2/3 trail. If the prints seemed disproportionately large, it was because the hominid's feet had slid a little in the rain-slick ash with each step it took.

A strong argument Mary and others cited against this interpretation was the presence of what appeared to be a heel mark in one of the G-2/3 prints a few millimeters inside the rim of another heel mark. Ron insisted that this feature should not be given undue weight, without explaining why. Not surprisingly, his protestations were largely ignored. Finally, at a conference held in 1985 in South Africa to honor Raymond Dart, Ron felt he had to reveal the truth: Among other damage done to the trail, Mary Leakey had inadvertently carved the heel mark into the print with her chisel.

"Although I saw she had dug through the calcareous lining of the print and advised her to stop," Ron told the assembly, "she did not heed my advice."

The audience was stunned. Before the session could be adjourned, Michael Day, a longtime colleague of the Leakeys and one of the few scientists who had actually seen the original prints, demanded to be recognized.

"I cannot let this session pass without responding to the outrageous suggestion made by Dr. Clarke," Professor Day angrily declared. "The suggestion that Mary Leakey was incompetent and damaged those footprints, while she is not here to defend herself, is one of the most outrageous things I've heard in a scientific meeting in many years." His censuring of Clarke's remarks was greeted with applause.

"Ron was in a bind," I explained to David and his friends. "He's by nature a mild-mannered guy with a deep respect for Mary's contributions. But he believed that there were really only two sets of prints on that site at Laetoli. To make his case, Ron had to say things about Mary's chiseling technique that weren't going to make him very popular."

"Most of us disagree with him about the number of hominids who walked here," Tim added. "But today, everyone knows that Ron Clarke was right about that artificial heel. And he deserves respect for having the courage to risk his own reputation and say something about it."

If Tim White bristles when he talks about the Laetoli excavation, it is only because he values the footprints themselves so highly. As evidence, they bear upon one of the most critical questions in the study of human evolution: why our earliest progenitors stood up and started walking on two legs. Between the discovery of Neandertal Man in 1856 and *A. afarensis* in the early 1970s, numerous fossil bones had paid testimony to our ancestors' ability to walk upright. But here at Laetoli, startlingly unforeseen, was even more direct testimony to the function that triggered the human career—the mark of heel and toe, muscle and skin, pushing down upon the ground.

No one questions anymore whether hominids before *Homo sapiens* walked erect. Until recently, however, most scientists were reluctant to believe that Neandertal Man, *Homo erectus,* and the rest really walked *like us.* Their style of locomotion has been variously described as "unique and distinctive," "not fully human," "inefficient," "waddling," and a "shambling half-run." At the heart of this reasoning is the seemingly obvious truism that before our forebears learned how to walk, they had to learn how to walk badly. In the opinion of some other anatomists, however, the people who talked about semihuman styles of bipedalism in the past had fallen into the old anthropocentric trap—the tendency to regard human evolution as a series of inevitable refinements, each of our ancestors becoming a little more refined, a little more like us than the one that had come before.

In the early 1970s, Owen Lovejoy of Kent State in Ohio set out to study the biomechanical properties of *Australopithecus* postcrania from South Africa to see how these early ancestors locomoted. Previously, the anatomy of these pelvic and limb bones had been used as evidence to show that *Australopithecus* was a rather clumsy walker. Working closely with orthopedic specialists, Owen demonstrated that in fact these ancient hominids were *better* walkers than we are today, owing largely to the subsequent compromises that *Homo* had to make in pelvic structure so that human females could accommodate the birth of large-brained young.

Owen went on to take charge of the postcrania that we were pulling out of Hadar—a mammoth responsibility. He subjected the bones to X rays and CT scans, and carefully weighed every anatomical nuance. Nothing that he saw in that wealth of bones shook his conviction: The earliest known hominids were fully committed to a bipedal, terrestrial life-style. Granted, the postcranial bones exhibited some "primitive" aspects. But one would expect that, since *afarensis* was the most ancient hominid. The arms appeared long, especially relative to the short lower limbs. The finger and toe bones were more curved than ours. But the "total morphological pattern"—the important suite of features associated with locomotion in the hip, knee, ankle, and foot—sang out strongly in favor of bipedalism.

For a while it seemed that Lovejoy had had the final word on the subject. But in 1983, two investigators at the State University of New York at Stony Brook announced that *their* study of the Hadar and Laetoli material convinced them that Lucy and her kin had in fact spent a significant amount of their time not walking on the ground but climbing in the trees. Granting that in most respects *afarensis* was bipedal, Randy Susman and Jack Stern detailed more than two dozen separate anatomical traits suggesting that the species was nevertheless a less efficient biped than modern humans. In their opinion, *afarensis* was no less than a "locomotor missing link." They even suggested that the females of the species were better adapted to climbing in trees than their male counterparts. The specter of a shambling, bent-kneed quasi-human was out of the closet again, this time backed by a great deal more evidence than had ever been available before.

Stern and Susman were not the only ones denying *afarensis* a two-legged locomotive perfection. The year before, their Stony Brook colleague Bill Jungers had argued that Lucy's legs were too short, in relation to her arms, for her species to have achieved a fully modern adaptation to bipedalism. At the same time, Russell Tuttle of the University of Chicago was pointing to the fact that the Hadar specimens possessed curved fingers and toes, possibly an apelike adaptation for grasping tree branches. Meanwhile Brigitte Senut and Christine Tardieu in Paris were arguing, on very thin evidence, that the Hadar knee and elbow joints also indicated a much more arboreal animal than Lovejoy's bipedal ideal.

Owen Lovejoy was not at all dismayed that Lucy's status as a full-blown biped was being attacked on several fronts.

"The first rule of anthropology," he told me, "is that when everybody believes what you've said, you've probably got it wrong."

Owen's imperturbability notwithstanding, I felt that what was needed was a forum bringing together all the people whose work was converging on the critical question of Lucy's locomotion. In the spring of 1983 the fledgling Institute of Human Origins sponsored its first scientific conference, "The Evolution of Human Locomotion." Tim and I got together and prepared a list of invitees. Everyone responded enthusiastically, except for the French investigators, who chose not to come. Tim himself agreed to talk about the Laetoli footprints. I would moderate the proceedings and try to make sure things did not get out of hand.

That's not always an easy assignment with Tim White around. When he took the podium early in the conference, Tim immediately challenged a specimen found by Noel Boaz, a scientist who was in the audience. In 1979, Boaz had found what he believed to be a hominid collarbone at a site called Sahabi in the Libyan desert. The deposits where the bone had been discovered had once been covered by a salty sea, and they could be dated by the presence of certain marine-plankton fossils to about five million years. That date could not be more significant: It made Boaz's fossilized clavicle the oldest known hominid in the world.

"The bone has a backward-S curve that is characteristic of hominids," Boaz had reported earlier, "implying that the creature may have been bipedal." If that was true, then hominids were walking erect fully two million years before Lucy.

Before the conference, Tim and his students had done a little study of Boaz's fossil, and now he dropped a bomb. Far from representing a hominid or any other primate, Tim told the audience, the specimen was the fossilized rib of a dolphin. The S-shaped curvature of the bone that Boaz had built his case around was simply not evident in the specimen. Boaz rose from his seat and angrily defended his position, ticking off a number of anatomical points that gave the bone "about a 70 or 80 percent chance" of being a hominid. But Tim was not there to play probability games.

"For those of you who question whether the specimen Dr. Boaz found in Libya is a dolphin or a hominid," he said from the stage, "we invite you to take a look at a dolphin rib and a cast of the Sahabi fossil up here after the talk. I don't think there is a single

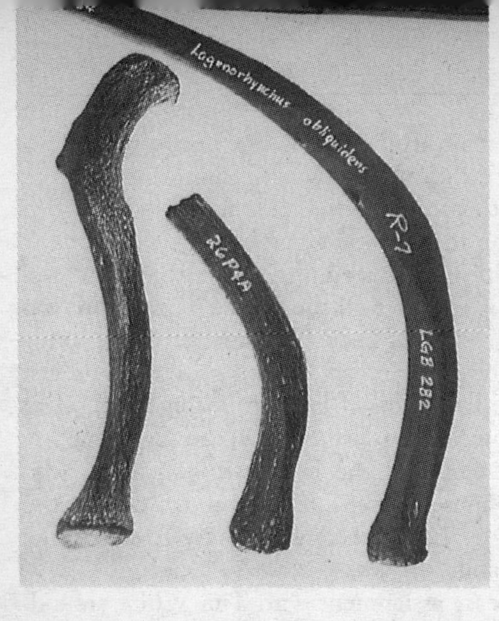

The Sahabi fossil (center), found by Noel Boaz and believed by him to be the clavicle of a five-million-year-old hominid. Boaz claimed that the fossi possessed an S-shaped curvature, like that of a chimp clavicle (left). Tim White disputed the claim, declaring that the specimen was the fossilized rib of a dolphin, like the one shown at right. TIM D. WHITE

person in this room—perhaps not even yourself, Noel—who will walk out still thinking that this bone is anything except a fossil dolphin." Later, Tim took to calling the fossil "Flipperpithecus boazi."

"Unseemly behavior," Boaz told a reporter. "It's just his style, so what can I say?" He later claimed that in making his point, Tim had engaged in *ad hominem* sophistry.

"In fact I argued with *ad delphinem* actuality," Tim retorted. A picture of the Sahabi specimen is above, flanked by a chimp clavicle on the left and a modern dolphin rib on the right. I'll let the reader decide how much of an S-shaped curve the fossil shows.

The conference then dove into the meat of the dispute over bipedalism in *afarensis.* Russ Tuttle discussed what he saw as an unresolvable discrepancy between the Laetoli footprints and the foot of *afarensis* that supposedly made them. From his analysis of the footprint casts, Russ concluded they had been left by a creature walking more or less like a human. But the foot skeleton we had found in the Hadar had long, curved toes and other features that in his view would have left a different sort of print in the ash. The implication was that an altogether different hominid species—the ancient true *Homo*—had existed at Laetoli and left its tracks behind. If that was correct, then it followed that, in our phylogeny, Tim and I had crammed more than one species under the name *afarensis.*

"You saw Tim White fracture Mr. Boaz's collarbone," Russ joked with the audience, "and I guess he's ready to stomp on my feet."

Tim did have his say later on. Tuttle had based his analysis on the best preserved of the Hadar feet—a large, nearly complete foot skeleton from the First Family site. Tim acknowledged that with its long, curved toes, this foot would not come close to fitting into the prints at Laetoli. But when Tim and Gen Suwa had scaled the anatomy of the specimen down to a Lucy-like size, the *afarensis* slid into the Laetoli prints with a Cinderella-like fit.

Randy Susman, bearded and intense, made the case at the conference for Lucy's climbing abilities. Susman was careful to stress that no single feature endowed *afarensis* with such ease in the trees, but rather a functional complex of traits working together. Chief among these in the upper body were strong, curved, apelike finger bones, and Lucy's long arms relative to her legs—another primitive trait that would facilitate movement among the branches. Below the waist, Susman pointed to what he and his colleagues believed were greater ranges of motion in Lucy's hip, knee, and ankle joints. Lucy also possessed those curved toes, and according to Susman, the anatomy of the base of her foot and her calf bone indicated that powerful muscles once stretched between them. Both of these traits would have given her a grasping prehensility in her feet that would make life a lot easier aloft. In sum, the creature conjured up by Susman's arguments would have been a fairly respectable arborealist, scampering along branch tops, swinging between limbs, and shinnying up trunks with an agility that would put our own attempts to shame.

Jack Stern then took the podium to present the evidence for a less-than-human mode of bipedalism in *afarensis.* He cited Bill Jungers's work on Lucy's short femur, and discussed some fine points of hip, knee, and ankle anatomy. But he leaned most heavily on what the Stony Brook team saw as a peculiarly primitive orientation in Lucy's pelvis. In apes the large flaring bones of the pelvis, called the iliac blades, are tall, narrow, and rather flat back to front. Human pelves are shorter and broader in the beam, and the iliac blades twist around to the side. Everyone agreed that Lucy's pelvis was short and broad. But according to Stern, it still lacked that crucially human twist to the side. This difference has everything to do with locomotion. Without that feature, Lucy could have kept her balance only if she walked with her hip joint flexed, like a chimp.

When it came to the Laetoli footprints, Jack agreed with Tim that *afarensis* was the species that had left them behind. But from the Stony Brook perspective, those Laetoli hominids were walking like missing links. When chimpanzees walk bipedally, they transfer their weight in each step along the outside of the foot. Humans carry their weight from the outside to the inside, so that the last, powerful push-off phase of a stride is supported by the big toe. The ball of the foot should therefore leave a deep impression, especially in the soft surface. But in many of the Laetoli prints, Stern and Susman could find no such deep impression. In one print they actually saw a swelling where there should have been an indentation. They concluded that at push-off, the Laetoli hominids at least sometimes kept their weight on the outside of the foot. Earlier, Stern and Susman had pointed out too that in at least one of the prints, the big toe appeared to be splayed out to the side instead of lined up with the other toes, as in the human condition. Perhaps, they argued, the Laetoli hominids were capable of abducting their big toes outward, almost like chimps.

The Stony Brook arguments about the Laetoli prints made a big impression on some people at the conference. But like almost everyone else, Randy and Jack had been basing their conclusions on studies of casts. Tim knew the originals in detail, and he made it abundantly clear how misleading casts can be. The print that Stern and Susman had pointed to as most obviously lacking an indentation turned out to be one of those excavated by Mary Leakey's chisel probe. Most of the fine detail was obscured. Tim also pointed out that when humans walk on soft surfaces, their tracks do not always show a well-defined imprint of the ball of the foot either. As for the splayed-out big toe, Tim showed that in the ash surrounding that print there were tracks of a Pliocene hare. As the hare hopped through the ash, it left one footstep overlapping the hominid print, creating the false impression of a big toe. This detail could be plainly seen only on the original footprint.

Stern and Susman later admitted that after the revelation that some of the footprints had been improperly excavated, they had "much reduced faith" in their conclusions about them. Instead, they threw their weight all the more behind their arguments from the bones from Hadar. But Owen Lovejoy was quietly waiting in the wings, and now it was his turn. He believed the Stony Brook team had missed seeing the forest while looking for Lucy in the trees. It wasn't enough to simply collect a few simian resemblances and declare a locomotor missing link.

"Bipedality demands a fundamental reorganization of the lower limb," Owen declared at the beginning of his talk. "The animal *has* to give up one form of adaptation for another. There is no such thing as an anatomical free lunch."

A case in point is the human pelvis. The only way to design an animal that can balance the weight of its spinal column in a vertical position is to give it a rounded bowl of a pelvis. But if that remodeling were to be effected on the narrow, keellike pelvis of an ape, the animal would have to sacrifice some 30 percent of its leaping ability—hardly what you would expect in a creature locomoting limb to limb. Stern and Susman had maintained that Lucy's pelvis lacked the human twist to the side, but again they had been deceived by casts. Lovejoy had spent long hours with the original, badly fragmented fossil, carefully restoring it to what he thought was close to its form in life.

Owen's restoration did not reveal a pelvis exactly like that of a modern human, but it did show a much more human like twist to the side in the iliac blades than Stern and Susman had contended. The crucial question was, Did Lucy's hip joint work like ours, or like that of an ape? Owen demonstrated convincingly that the abductors—the hip muscles—employed in stabilizing the hip in bipeds operated in Lucy just as they did in modern humans. He also showed that Lucy's abductors assisted in eliminating potentially dangerous stress upon her *femoral neck*—the thin bridge of bone connecting the shaft of the thighbone to the balllike femoral head, which fits into the hip socket.

A closer look at this vulnerable stretch of bone revealed a telling difference between bipedal humans and quadruapedal apes. An arboreal ape subjects its femoral neck to extreme bending stresses when it leaps and climbs, resulting in a thickening of the bone at the top and the bottom. In a human femur, the femoral neck is adapted only to stresses imposed by walking and running—resulting in a thickening of bone just along the bottom, with only a thin layer of bone at the top. Lucy's femoral neck is identical to that of a modern human, and would fracture if subjected to high stress during climbing.

"Lucy was not just capable of walking upright," Owen later wrote. "*It had become her only choice.*" Point by point, Owen and his student Bruce Latimer showed how, below the waist, Lucy was factory-built for full bipedality. One of the most important bits of evidence that Stern and Susman were using was those curved *afarensis* toes. But as Latimer made clear, in the perspective of the whole functional anatomy of the foot, the toe issue withered in importance. He pointed

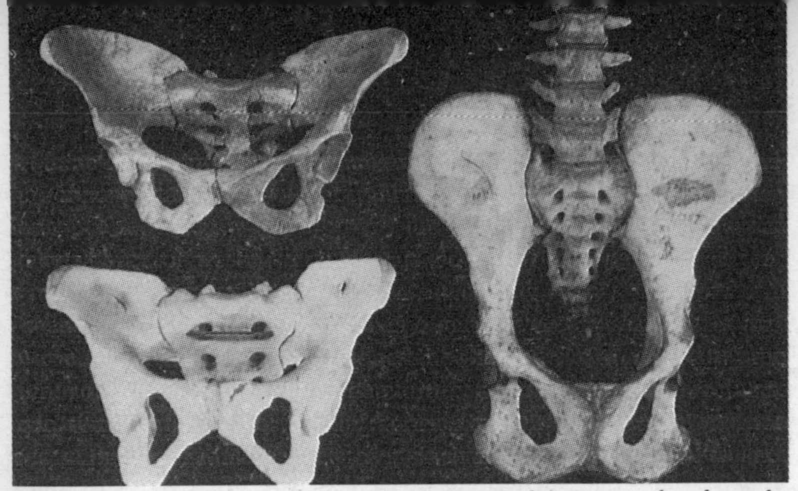

Owen Lovejoy's reconstruction of Lucy's pelvis (upper left) compared to those of a human (lower left) and a chimpanzee (right). Lovejoy maintains that Lucy's human-like pelvis is one important indication that her species was fully adapted to bipedal locomotion. C. OWEN LOVEJOY

out that the foot is the most distinctly human part of our anatomy: We have lost the opposable big toe, the prehensile grasping action of ape feet, and gained instead the springlike arch and rigidity of the forefoot needed for bipedal walking. And contrary to the Stony Brook line of reasoning, the *afarensis* ankle, knee, and hip joints were unquestionably human in all but the most trivial respects.

Whichever side one took in this debate, the formal presentations, technically complex but passionately argued, had brought the issue to a fever pitch. On the morning of the second day we brought everyone back together for a closed-door session in the presence of the evidence alone. With no set agenda and no press allowed, the disputants were free to let loose without fear of public scrutiny or the temptation for public grandstanding. Quickly small groups coalesced in pockets around the room. When I looked up an hour later, Tim White, Gen Suwa, and Russ Tuttle were burrowed in one corner, fitting and refitting casts of toe bones into the Laetoli prints. Bill Jungers was running from table to table, measuring bone lengths and ratios. Owen Lovejoy and Jack Stern were pointing finger fossils at each other. Bruce Latimer was explaining ankle esoterics, and Randy Susman had bared one foot, the better to demonstrate toe curling. There were raised voices and arms being thrown up in exasperation, but there was also a lot of listening going on. I smiled to myself. This is why I had put the Institute together. All the work had paid off.

By the end of the conference, I was convinced that Lovejoy and

Co. had made a convincing case, at least with respect to anatomy below the waist. When it came to upper-body adaptations, I was not quite so sure. What bothered me were Lucy's long arms and strong, curved fingers. Both would be a big advantage in moving about in the trees, regardless of whatever commitments a species had made toward bipedality from the waist down. On the other hand, Lucy's long arms and curved fingers might be examples of what Lovejoy calls "evolutionary baggage"—traits left over from a more distant ancestor, not yet lost but no longer needed. I, for one, cannot imagine long arms somehow being a *disadvantage* to a fully bipedal hominid.

Stern and Susman notwithstanding, it seemed to me that Lucy's lineage had already crossed the line over to bipedalism, never to return. I put the question in the back of mind and went on to other things. On that expedition to Olduvai, five years later, it would all come back again.

David Koch had arranged for a chartered plane to fly in from Manyara, pick up his safari, and take it over the Kenyan border to the game-rich plains of Masai Mara. But the soft sand of the makeshift Olduvai landing strip was too perilous for the little Cessna that flew in the next day. After nearly crashing on landing, the pilot refused to burden his take off with passengers. The safari would have to "rough it" back to Manyara by Land-Rover. The decision was unpopular, but it proved wise. Carrying only the pilot, the Cessna barely cleared the runway on takeoff, practically scratching its belly on the thorn fence surrounding the camp.

The near-miss put an end to any grumbling about the bumpy ride back to Manyara. David and his friends crammed into the Land-Rovers and headed out of the Gorge. An hour later another A&K vehicle churned up the road into camp, bringing supplies. The Binfords and the Frisons had been waiting for an opportunity to move on too. While the A&K driver paced impatiently, bags were hastily packed and thrown into the rear. The party was over. It was time to get back to work.

On Dik-dik Hill it was the old story. The hominid hinted at its identity, promised meaning, but refused to deliver. We scalped the top off the hill, hoping that the little knoll might have held on to some bits of the skeleton even as the greater part of it eroded out and washed away. We were rewarded with a single molar crown fragment, not enough to tell whether the hominid was *Australopithecus boisei* or *Homo habilis.* The most likely areas now exhausted, we started in on the longer shots, opening up a new line of squares farther down toward the road. For our efforts we gained a few more tooth roots, a splinter of crown, a handful of brown-black enigmas. By this time the mound of backfill had grown so large we could rest up against it while we worked the karais. There was still much to be done, and in each fresh square there was still the hope of the big find we had imagined from the beginning. But the chances of that happening were tapering away by the hour. Each of us knew, alone, that the best of it was over.

Gradually, imperceptibly, the locus of our expectations shifted toward the lab. On a table by a well-shaded window, we laid out the labeled plastic bags stuffed with bone, dark splinters of possibility. The sheer anonymity of those bones was daunting—thousands of them, of which perhaps one in a hundred really belonged to the hominid, and only one in a hundred of those could ever be recognized for what it was. Yet somehow they all had to be sorted, tested, and worried into order.

The familiar comparison between doing a jigsaw puzzle and reconstructing a fragmented fossil is accurate only up to a point. Yes, the process is much the same—pick up a piece, scan the possible points of connection, test edge against edge until finally something fits. But keep in mind: The vast majority of the pieces in a "fossil puzzle" have been lost. Among those that remain, many have lost their defining edges. In the pile of fragments confronting you, most belong to an an unknown number of *other*, largely irrelevant puzzles. At most, you'll succeed in re-creating only a tiny fraction of the picture you want to see. But if you mess up, a priceless opportunity is gone, forever.

In spite of such difficulties, or perhaps because of them, that table by the window in the cool stone building became an irresistible lure. July turned to August and Dik-dik Hill became a sort of distracting obligation, a chore that took us away from the *real* work. As soon as there was light to see by in the morning, somebody would be over in the lab, pinching a couple of fragments together, rolling them around between thumb and forefinger, searching for joins. The chances of success with each attempt were grotesquely remote, but that seemed only to sharpen the compulsion. After breakfast we'd pack up to go down to the Hill, but if somebody was late getting ready or had to run back to their hut to fetch a canteen, somebody else would slip unobtrusively back into the lab and work the bones for a minute or two. It was the same during the afternoon break, or in the hour before dinner—steal away to the lab and try some joins. In the evening we got out the flashlights and worked until our eyes lost their focus.

Once, in the middle of the night, I got up to relieve myself outside my hut. The lab building was a dense black absence, its shape defined against a horizon built of starlight. A thin halogen glow leaked from the lab window. Somebody was at the worktable, bothering the fossil under the beam of a Mini-Mag flashlight.

People love to hear about breakthroughs, about sudden transformations in understanding that turn ignorance to knowledge, base metal into gold. Science is rarely so dramatic. Before that expedition, however, I would have said that the process of discovery always leaves behind recognizable landmarks, points along the way where outstanding questions were answered, doubts satisfied. Now I'm not so sure. I can easily reconstruct the trail of clues that the hominid left for us, and when we came to solve them. It is all duly recorded in my journal. But what the journal lacks is the one entry you would most expect to find: "Eureka! Today we solved the identity of the hominid!" That entry isn't there because it did not happen that way. There was no eureka. There was no grand turning point. The evidence kept dribbling in, and through hard labor and some dogged thinking we *did* solve the puzzle, not through revelation but through a sort of absorption, just below the level of explicit conciousness. It was as if the truth had slowly seeped in through our pores, until we had come to know it without knowing that we did. So when the final, indisputable confirmation came, we hardly noticed the event. What had once been a mystery had become—in hindsight, mind you—obvious from the start.

Our hope for pinning down the taxon of the hominid had come to rest mostly on Gen Suwa's shoulders. Mary Leakey had taken with her the lab's collection of casts that would have helped in making comparisons, but Gen—deep into his dissertation on hominid dentition—carried a lot of what we needed inside his head. Whenever there was enough light and a little time, Gen could be found squinting over the collection of tooth fragments, bringing his memory to bear on some groove barely evident in a chip of crown, imagining its extension and searching for that image in some other piece.

Once again, the nature of the fossil thwarted progress. The Dik-dik Hill hominid was old when it died. ("A real Mzee," said Tim.) Most of the surface features of the tooth-crown fragments had been worn away. Our best chance of discriminating between *boisei* and *habilis* lay in finding an intact third molar. Like a human wisdom tooth, a third molar would have erupted late in life, and consequently in death might still retain enough of its surface topography to make a positive ID. But just knowing the size and dimensions of the tooth alone might be enough to tip the taxonomic balance. I'll say it again: Robust autralopithecine cheek teeth are huge, relative to the size of their canines and incisors. *Homo* cheek teeth are proportionately small. To make the discrimination, you need one front tooth from your fossil, and one cheek tooth. We already had that canine from the dik-dik pile.

Meanwhile, the fossil's postcranial skeleton was slowly taking on shape. Shortly after Koch's departure we unearthed another inch or so of humeral shaft. When Gerry Eck and I came into the lab one afternoon Tim had just succeeded in grafting the piece on to the existing length of bone. The effect was chilling. We now had virtually the entire humerus save for the articular ends. A peculiarity of the bone, merely hinted at before, was now baldly obvious, laid out on a strip of cotton with the transparency of fact.

"Step right up, folks, and take a peek at the ape lady," Tim said, pushing his chair back from the table.

"Jesus. That is one *long* mother," Gerry said.

"I'll bet you that bone's even longer than Lucy's humerus," I said.

"But what we've got of the femur looks so damn small," said Gerry.

"Right." Tim reached over and picked up the chunk of calf bone we'd found the first day at the site. "But I can't make out this tibia. I just don't see how a tibia that size goes on to that little femur."

"Are you saying it might not be hominid?"

"I'm saying it might not be the *same* hominid. That's been my fear all along. That we might be digging around and suddenly come up with two teeth—a *habilis* incisor, and a great big *boisei* molar."

"Now look who's worrying too much," I said. "Every scrap of tooth crown we've found is worn down to the same degree. And we've got no duplication of any one piece of anatomy. That's enough for me to believe there's only one hominid out there."

"Okay, I'll buy that," Tim replied.

"So we've got a creature smaller than Lucy, but with longer arms in relation to its legs. Just as apelike, but a million years more recent in time!"

"Maybe."

"But that would put hominids in the trees long after . . ."

"Not necessarily. Let other people jump to whatever conclusions they want to. All I'm saying at this point is that this *might* be a very short hominid with unusually long arms. We're going to have to find more of the femur first to nail it down."

I didn't answer. Along with an intact third molar, the rest of that femur, especially the distal end, would be the most exciting thing Tim could turn up with his rock hammer. We had every right to hope for it. The articular end of a femur is just about the hardest piece of bone in a skeleton; you could drive a truck over one without doing much damage. But this fossil had denied our expectations. Since the very first day of surface excavation, there had been no more trace of the femur.

Then, on the first Saturday in August, a possible piece of limb shaft appeared in somebody's karai. When we returned for lunch I dropped my things and went over to the lab. Tim was there, intense as usual. Gen was immersed in the tooth roots. I picked up the morning's find and the portion of femoral shaft we had already glued together. They were the same color and texture. I turned the new bone around and clicked it down on the bottom of the shaft.

"Son of a bitch," I said.

"Let me see that." Tim reached out to grab the bone.

"Wait a minute!" I said, pulling away. I wasn't sure yet of the fit. Sometimes the match between two bones can look good on the outside, especially if you really *want* it to, but when you look inside, the interior edges don't come together. I turned the bones around. They lined up perfectly inside. The new piece added an inch to the length of the shaft: a critical inch more toward the distal end.

"The best part of this sucker is the break on the far end," Tim said when he finally got his hands on it. "Look how fresh that break is. The distal end has *got* to be out there."

Now the fossil was beginning to look like a skeleton, especially in the arms. We had most of the humerus. We had almost all of the radius too, and a solid stretch of ulna. If we *could* find the distal femur, we would be able to accurately estimate the hominid's body size. We could also calculate the length of the humerus relative to the femur. That would be a crucial statistic. In modern *Homo sapiens*, the humerus is approximately 75 percent as long as the femur. Apes have arms longer than their legs, so their "humeral-femoral index" is over 100 percent. We knew that Lucy's was in between, about 85 percent. Where along that spectrum did the new hominid lie? It was commonly assumed that when a younger hominid fossil like this one was discovered, it would have a humeral-femoral index closer to our own. But maybe that assumption was about to go by the board.

Bill Kimbel was due to arrive a couple of days later, and if he had received our telegram he would have Lucy's casts in his duffel bag. On the morning of August 5 we heard a Land-Rover whining in low gear down the far side of the Gorge. It was a Department of Antiquities vehicle, with Pelaji Kyauka, the young director of the Arusha Museum, at the wheel. On board were Fidel Masao and Alberto Angela, an Italian graduate student representing the Ligabue Research and Study Center, who would be helping us with the dig. Masao had with him the permit to survey at Laetoli. They told us that Kimbel and Paul Manega were following in another vehicle, having stopped at Ngorongoro for supplies. An hour later the second Land-Rover pulled into camp. We rushed out to greet it. Bill had received the telegram. He had the casts. He also had the two pounds of peanut butter.

"Things are brewing, Bill," I said as we hurried over to the lab. We crowded around the worktable and Kimbel laid out Lucy's limbs. Then we set each of the new hominid's bones up against the corresponding cast of the famous *afarensis* skeleton.

"Look at that. Holy shit."

"The femur's *smaller*. Smaller than Lucy."

"We thought so. But we couldn't convince ourselves. But check out the humerus. It's *bigger* than Lucy's!"

At that point, we couldn't tell the exact dimensions, the precise proportions. That would require more study back home, and with luck, more bone from the site. But we knew the gist of the story the moment Kimbel laid down those casts. The little hominid was almost certainly female—if it were male, then the females of the species would have been only two feet tall or so, which seemed impossible. More important, the skeleton showed the body proportions of a very primitive australopithecine. Yet it had lived in a time much closer to ourselves, a time when tools were already being used and discarded on the ground, a time when a being with brains and culture enough to be called *Homo* lived at Olduvai Gorge.

And we knew something else. Forgotten in the obsession about finding the rest of the femur was Gen Suwa, patiently fingering his tooth fragments, searching for the fits. Just before Bill's arrival he had succeeded in piecing together much of a third molar. The tooth was big, but not *that* big. It was the confirmation we had sought, and no longer needed. Bill, an expert on hominid-skull architecture, could read the same message in the fossil's face. This little creature, fully mature but standing somewhere around three feet three in its bare bipedal feet, was the same species as other fossils from Olduvai like George Twiggy and Johnny's Child, and most likely the same species as the famous big-brained 1470 from Koobi Fora. The same species that Louis Leakey had found, named, and argued for through the 1960s. This hominid, with a body like Lucy but presumably a brain empowered far beyond her reach, was *Homo habilis.*

Lucy's child. *Homo.* One of us.

That evening Gerry Eck reached into his secret larder and pulled out another few bottles of Tanzanian beer, dark and perfumy. We had cause to celebrate. Good friends, old and new, had arrived in camp. They had brought along an assortment of culinary riches to soothe our goat-numbed palates. Chickens. Fresh eggs and cheese. Peanut butter. M&M's. There was new hope for Laetoli, if we could convince Kyauka to stay on and lend his vehicle to the cause. With the fresh muscle power provided by Bill and Alberto, there was even some hope that the next day Dik-dik Hill might yield more than its meager daily ration of bone chips. Even if it didn't, we had already proved the enduring fertility of the Olduvai beds. We had found the first skeleton known of the earliest species of *Homo.* I was well aware, of course, what such superlatives meant for the future of the Olduvai research project, and the Institute of Human Origins as a whole. Understanding the complexities of paleoanthropology can be taxing to potential supporters, people who have important other concerns and little time to spare. Words like "first" and "earliest" cut to the quick.

But what struck me most of all about the skeleton was its perplexing primitiveness. Most experts on hominid body size would have predicted that a skeleton of *Homo habilis,* when found, would prove to have a stature and body proportions somewhere between those of its *afarensis* ancestor and its descendent *Homo erectus.* Luckily, we had skeletons of both those species to refer to. Lucy, a female *afarensis,* stood about three and a half feet tall. In 1984, Kamoya Kimeu had made a sensational discovery at West Turkana; a nearly complete skeleton of a twelve-year-old *erectus* boy. The skeleton, found in deposits dated at 1.6-million-years old, stood about five feet four inches tall. If one assumes human rates of growth, the boy would have reached six feet in height as an adult—as tall as a modern American male.

If body height in the human line did indeed increase gradually from *afarensis* to *erectus,* then by rights *Homo habilis* should have averaged somewhere between four and a half and five feet tall. Instead, we had found a *habilis* skeleton that appeared to have stood no taller in life than Lucy herself. Judging from the fragments we had of the Dik-dik Hill hominid, from the neck down, she was practically Lucy's twin.

"What an odd luck of the draw," I said to Tim as I turned into my sleeping bag that night, "a diminutive specimen for *Homo habilis,* and a relative giant for *Homo erectus,* only two hundred thousand years later."

Neither we nor anyone else could tell, of course, how representative the new skeleton was of *habilis* female generally, or indeed whether the Turkana boy was typical of *erectus* males. But from the admittedly scant information we now had in hand, and making the fair assumption that *Homo* was no more sexually dimorphic than *afarensis,* then in a span of not more than two hundred thousand years, our ancestors had made a surprising leap in body size. Even more intriguing was the abrupt change in body proportions. The Dik-dik Hill hominid had long arms like Lucy, but *Homo erectus* had upper and lower limbs proportioned more like our own.

A whole stew of ideas was simmering in my mind that night. We already knew that *Homo erectus* brains were significantly bigger than those of *habilis*—and now it seemed evident that their bodies were qualitatively different too. Something extraordinary was happening during those few hundred thousand years of evolution, some shift in behavior that would quite suddenly transform a *Homo* with a comparatively small brain in a primitive body into one with a big brain and a modern body. That *Homo erectus* put that larger brain to good use is evident throughout the archeological record. Its material symbol is the beautiful, biface Acheulean hand ax, an enormous technical advance over the rough cobbles associated with *Homo habilis.* The discoveries of *Homo erectus* fossils all over the Old World testify to the species's ability to solve the complexities of living in colder climates—arid steppes, temperate woodlands, high-altitude plateaus. According to some investigators, the species knew the use of fire, perhaps even language. Indisputably, it was the first hominid species to venture out of Africa—not just to dip a toe into neighboring regions of Europe and the Middle East, but to fling its presence northward and eastward into modern Hungary, Germany, and the far northwestern regions of China. The significance of this vast radiation cannot be underestimated. If *habilis* was a local success, *Homo erectus* was an international phenom.

That's what really thrilled me about the Dik-dik Hill hominid. We now knew that a very primitive creature—from the neck down, little more than a walking ape—was practically rubbing shoulders in the early Pleistocene with a much more modern-looking descendant. Such a profound evolutionary acceleration does not occur without reason. Whatever the cause, an unprecedented evolutionary event had taken place, one that would hold ultimate consequence for the entire subsequent history of the planet. That night again I could not sleep, but it was not for anxiety. I was rethinking the old insights into that mystery, and the new possibilities.

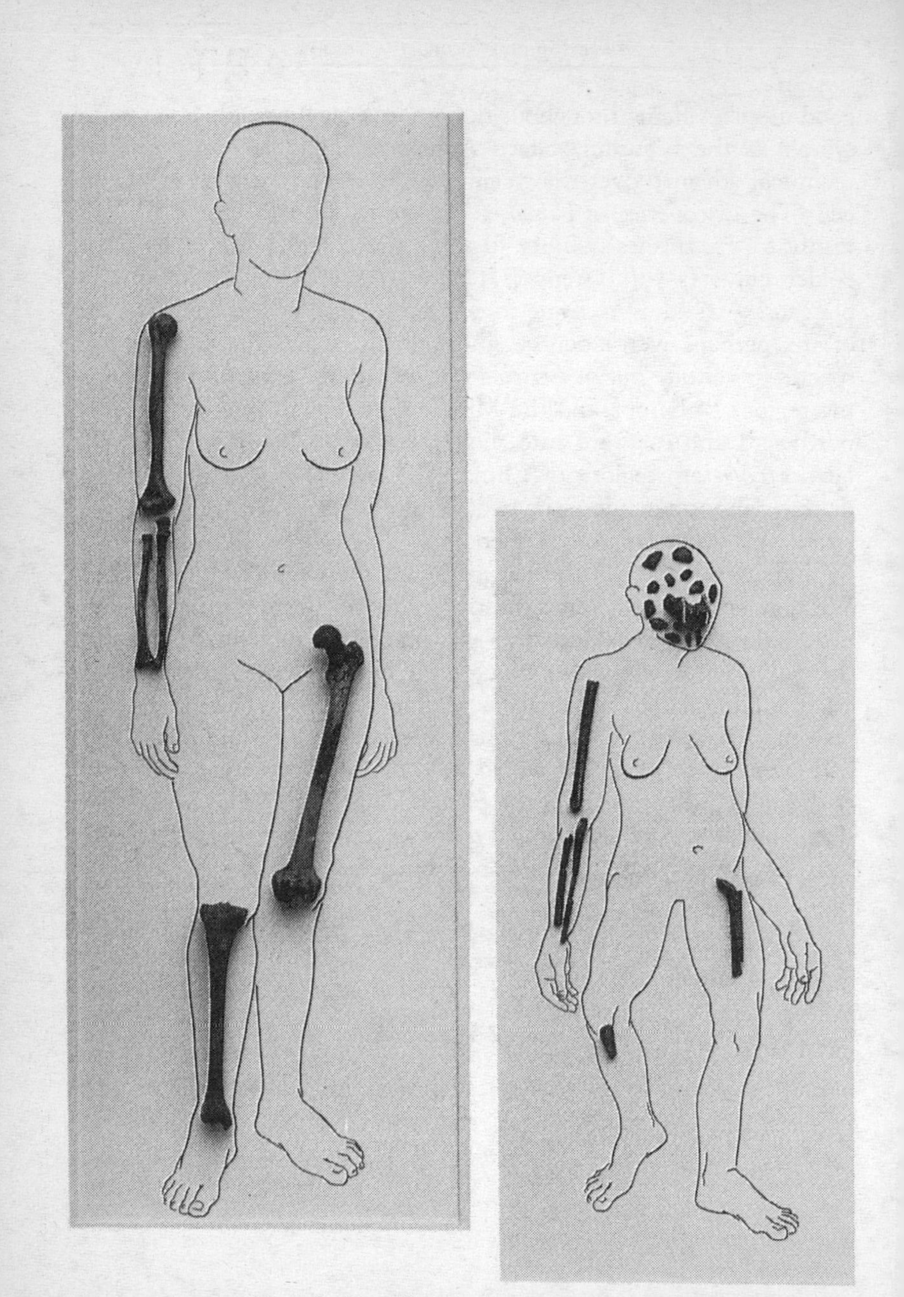

A reconstruction of the Dik-dik Hill hominid is compared to a modern human female. The skeleton found at Olduvai Gorge in 1986 would have stood less than three and a half feet tall. Her long arms and small stature suggest that the earliest human species was much more primitive in body size and proportions than thought.

PART IV:

Becoming Human

CHAPTER EIGHT: *What Happened at FLK?*

All the facts gleaned from the [Olduvai] deposits interpreted as living sites have served as the basis for making up "just-so stories" about our hominid past.

—Lewis Binford,
Bones: Ancient Men and Modern Myths

Because of a variety of weaknesses in Binford's analysis, it is doubtful that his results on Olduvai are a useful contribution to scientific archeology.

—Henry Bunn, in a review of *Bones*

This whole book review is an ad hominem *argument effected through innuendo and self-serving misrepresentation.*

—Lewis Binford, "A Reply to Bunn"

Hamana nale kui,
nale kui.

Here we go round,
go round.
—A Hadza dance

IF YOU WERE TO ASK a dozen paleoanthropologists at random to name the three most important letters of the alphabet, eleven of them would probably answer "FLK." The *Zinjanthropus* skull was found at

Frida Leakey Karongo, of course, and for that alone the site deserves the concrete plinth that stands there today. A year later the Leakeys turned up those first traces of *Homo habilis* nearby, and soon after the two leg bones practically right under Zinj's nose. But FLK yielded still more: a litter of mammal bones and stone tools so evocatively dense as to all but overshadow the hominid fossils themselves. If there is any one place on earth to ask the Big Questions—How did we begin? What were we doing two million years ago that made us human?—FLK is the spot.

Sometimes after a day on Dik-dik Hill I chose to forgo the Land-Rover and walk back to camp, a route that happened to take me right by the famous site. I remember passing once an hour before sunset. I was alone, and as I turned to face FLK, a robust, coppery light struck the back wall of the old excavation, flooding the intervening space. Mary Leakey's insistence on neat angles in excavations had produced walls so smooth and straight they looked as if they'd been cut into the hillside by a giant cake knife. The floor too was worn smooth. Of course, the thousands of ancient bones and tools that once graced it had long since been swept off to the Kenya National Museums in Nairobi. The only bones there now were the bleached remains of a wildebeest killed by predators only a few weeks before. Lit up by that sudden infusion of sunlight, and with the surrounding slopes and gullies already dimming in shadow, FLK seemed to me like a stage set for some minimalist drama: bare, austere, its props pared down to ephemeral meanings. At stage center rested the monument to Zinj, the marble plaque cracked, the concrete base weathered smooth. As a fixture on the set it would outlast the poor wildebeest, but not by much.

In a sense, I thought, FLK *has* functioned as a sort of theater. I sat down on a mound of compacted backfill opposite the site—a good seat in the house. There was no sound but the faint rustling of lizards and insects in the gravel around me, and the whisper of the breeze threading through the sisal on the hill above. Something occurred on this spot almost two million years ago, some events of no particular importance at the time, except that they involved creatures who would later become *us*. Even so, the events would have lived anonymously in the past forever, had they not left behind an indelible trace of themselves, and had not Louis and Mary Leakey rediscovered that record thirty years ago. Ever since then, what happened at FLK has inspired a series of scenarios attempting to replay those moments in the early Pleistocene, and in so doing explain our beginnings to ourselves. There are other archeological sites, of course, that have given substance to our interpretations of the past—many right here in the Gorge. But for a scenario to have credibility at all, it has to play at FLK.

First a prologue, produced by an old familiar. Before the Leakeys struck it rich in Olduvai Gorge, the prevailing view of our ancestors' habits was an extraordinary vision being put forth by Raymond Dart. In the years surrounding World War II, Dart had returned to the anthropological fieldwork he had abandoned following his disappointing journey to England to show off his Taung child. Instead of looking for hominid fossils, however, he was now searching for signs of hominid *behavior*—evidence of tools, the use of fire, or any peculiar pattern in the concentrations of mammal bones in the South African caves that suggested they had not come there through natural means.

Dart found his evidence in a lime quarry in the Makapan Valley in the Transvaal. There he and his students found several specimens of *Australopithecus africanus*. But more intriguing than the hominids themselves were some dark stains in the deposits, which Dart took to be the traces of ancient hearths. There was also a curious anomaly in the collection of seven thousand mammal bones also recovered from the site. When he tallied up the numbers of anatomical parts in the collection, Dart found that he had many more skulls, mandibles, and lower legs than could be accounted for by the number of ribs, vertebrae, and pelves also present.

Very curious. When an animal dies, all the parts of its skeleton will of course be present just as they were in life—so many ribs, limbs, and so forth. If you find a collection of animal fossils where there are more of *some* body parts represented than you would expect to find judging by the number of other body parts counted, then you have to assume that something has tampered with the animal's bones after it died. There are all kinds of forces that can influence the fate of a bone after death—scavengers can carry off parts of a carcass and leave others behind, for instance, or gnaw them to splinters, or water action over time can wash some smaller bones away from their original location, leaving larger ones behind. There is even a name for the subdiscipline of paleontology that deals with the forces acting upon bones after death—we call it taphonomy.

Dart reasoned that the force responsible for the balanced number of anatomical parts at Makapansgat was the hand of our ancestors. The most dramatic evidence for hominid involvement was a collection of forty-two crushed baboon skulls, fully two thirds of which had been smashed in on the left side. Dart believed that the baboons had all been slaughtered by right-handed australopithecine hunters wielding lethal clubs.

These weapons were manufactured of bone and other hard body parts—which explained the plethora of some kinds of bones at the site. He called the culture that produced them the *osteodontokeratic*, meaning "bone-tooth-horn," and painted a vivid portrait of its members:

"Man's predecessors," he wrote in 1957, "seized living quarries by violence, battered them to death, tore apart their broken bodies, dismembered them limb from limb, slaking their ravenous thirst with the hot blood of victims and greedily devouring livid writhing flesh."

With Dart serving up such high-protein prose all by himself, his "killer ape" hypothesis, as it came to be called, hardly needed a professional writer to escort it out of the scientific literature and into the public domain. As it happened, an American journalist and playwright named Robert Ardrey met Dart on a tour of Africa around this time, became enamored with the professor's osteodontokeratic bad guys, and wrote a best seller about them called *African Genesis*. Ardrey's writings, translated into many languages, spread Dart's word around the world. Possessed with a deeply cynical mind himself, Ardrey used Dart's grim vision of man's beginnings as a key to understanding—or even excusing—mankind's persistent tendency toward violence:

"Man is a predator whose natural instinct is to kill with a weapon," Ardrey wrote. "The sudden addition of the enlarged brain to the equipment of an armed already-successful predatory animal created . . . the human being."

Compared to that brute ancestor, the version of humanity that Louis Leakey was simultaneously re-creating here by the Olduvai lake seems positively dainty. Spurred on by the discovery of Zinj, Louis and Mary had spent the rest of the 1959 season digging further into FLK. On the same layer of silty green clay where Zinj had rested, they soon uncovered an interesting assortment of crude tools, and a litter of bones as well—bird's eggs, little birds, tortoises, and small, immature mammals. The complexion of this "faunal assemblage" fit neatly with Leakey's belief that the earliest toolmaking ancestor of Olduvai Gorge was essentially a vegetarian who ate meat only when he happened upon it. Certainly nothing about Zinj himself, with his big millstone teeth, contradicted this portrait of his life-style.

With the aid of the National Geographic Society, the Leakeys expanded the excavation the following year on a scale they could only have dreamed of before, exposing over three thousand square feet in their effort to find more of Zinj. The rest of him never turned up, but in the process, the faunal remains uncovered on the same level took on a compelling new character. Along with the scraps of tortoise, catfish, and lizards, the Leakeys found the bones of more impressive prey densely intermingled with stone tools: Pleistocene pig, horse, antelope, and giraffe. Suddenly Dart's killer ape did not look so out of place in East Africa, especially with the larger-brained *Homo habilis* specimens turning up to steal the lead role from *Australopithecus*. One of the *habilis* juvenile skull bones that had turned up nearby showed a curious fracture pattern. Leakey himself suggested—though he later regretted it—that the fracture might have been caused by "a blow from a blunt instrument." Other researchers speculated that "true *Homo*" might also have killed and eaten Zinj too.

Now, with the shadows gathering in the gullies around me, I imagined a band of Dart's deadly apes running amok on my FLK stage, like demons in a Bosch landscape. This image of man continued to enjoy immense popularity through the 1960s, reaching Hollywood stardom in Stanley Kubrick's film *2001: A Space Odyssey*. But at the same time, scientists were beginning to ask some pertinent questions. Could these small early hominids really perform such prodigious feats of carnage as Dart had detailed? Couldn't some other agents be responsible for the smashed skulls and peculiar patterns of bones found at Makapansgat?

In the mid 1960s, a young geologist named Bob Brain decided to look a bit closer at *all* the taphonomic forces that could have influenced the accumulation of bones-not only at Makapansgat but at the key South African sites of Sterkfontein, Swartkrans, and Kroomdrai as well. Brain's work was a masterpiece of prehistorical detective work. He noticed, for instance, that the trees on the savanna landscape in South Africa did not grow stunted and scattered as one might expect in such a rain-parched habitat, but instead grew in clumps, with many of the trees surprisingly large. Brain found that their roots had tapped into isolated reservoirs of ground-water formed in natural fissures had developed beneath the trees, exposing the subterranean caverns to the open air. All sorts of debris from the surface could fall into these fissures—including bones, especially if carcasses were being eaten in the trees above.

Brain knew that the leopard is the only living predator that regularly eats its prey in trees.

He analyzed the makeup of the bones accumulating beneath trees attractive to leopards, and discovered that the number of anatomical parts included far more lower limb bones and skulls than ribs and vertebrae. This was much the same pattern that Dart had found in the supposed leftover meals of his killer apes. The skulls of the leopards' victims too were often smashed in, the same as the Makapansgat baboons. In *African Genesis*, Robert Ardrey had also cited as evidence of a case of "intentional armed assault" a juvenile australopithecine skull bearing two curious puncture wounds. In the course of his study Brain took a pair of calipers and measured the distance between the two puncture wounds on that skull. The gap matched the space between a leopard's lower canines. Far from being the chief consumer on the ancient South African savanna, *Australopithecus* appeared to be merely one of the consumed.

Brain's work did not end with leopards. He studied other animals who might have contributed to the contents of a cave over its life history. To better understand how scavengers might affect the pattern of bones in ancient sites, he undertook a study of the Khoisan pastoralists in Namibia—or rather, a study of their garbage. Khoisan raise goats, which they regularly butcher and eat in their villages. They also keep dogs, who are naturally accustomed to chewing up, carting around and otherwise leaving their imprint on the remains of their masters' goat meals. Brain found that after the dogs had finished their scavenging, the litter of goat bones left behind also resembled what Dart had found at Makapansgat. Apparently, scavengers had contributed their part to the complexion of the cave deposits. Thus the lives and deaths of leopards, owls, porcupines, hyenas, and yes, hominids too—the imprint of the whole ecosystem through thousands of years—was written in the record of bones in a limestone cave, as long as it remained open to the surface. Eventually the opening would collapse under its own weight or fill up with new deposits, sealing the record until a lime quarryman's dynamite brought it all back to light.

• • •

C. K. Brain's studies of South African cave sites helped put to rest the theory of "the killer ape." The accumulations of mammal bones—including hominids—were more likely the remains of prey devoured by leopards in trees above natural sinkholes. Puncture wounds in the skull of one juvenile australopithecine, once thought to be evidence of murder among ancient ancestors, proved to match the spacing between a leopard's lower canine teeth. ILLUSTRATION BY DOUGLAS BECKNER

Bob Brains's work confirmed something about the killer ape that others were beginning to suspect: The hypothesis may have had as much to do with recent events in the quarryman's world as with human behavior in the early Pleistocene. Dart's view of our beginnings was heavily colored by post war gloom. In the 1950s, the self-image of our society was aching from the revelations of its darkest potentials, and just coming to terms with a new, nuclear *Homo.* In a sense, the killing instinct—"this common bloodlust differentiator, this predaceous habit, this mark of Cain"—that Dart planted in our origins made it all make sense: No one was at fault, except humanity itself.

Certainly the Pleisto-drama that replaced Dart's scenario had a sunnier plot, as well as more tangible evidence to lend it credibility. Right here on the Zinj floor, the concentration of stone tools that the Leakeys found was far more solid testimony to the activities of early hominids than Dart's imagined bone weapons. There was also a telling pattern in the way the stones and bones were distributed across the site. Small flake tools and thoroughly smashed-up bones dominated a central area of about four hundred square feet. A semicircle of barren ground bordered this concentration on its eastern edge. Beyond the barren patch, the debris thickened again, but was characterized here by more complete bones, and larger tools. Desmond Clark visited the site and made the inspired suggestion that the quarter-moon of cleared ground might represent the former location of a windbreak—perhaps some sort of thorn fence—protecting the central living area of the tool-using hominids of FLK. Beyond the windbreak lay the debris chucked over the fence by the hominid occupants.

That the Zinj level at FLK represented an ancient "living floor," to use Mary Leakey's term, seemed self-evident; the only questions was what the residents were doing for a living. The many large animals represented on the site had all but convinced most scientists, including Louis Leakey, that the earliest Olduvai toolmaker was doing more to procure meat than just aimlessly stumble around tripping over lizards and baby gazelles. More evidence was soon unearthed: a disarticulated elephant skeleton found in upper Bed I deposits at the nearby site FLK N, and a skeleton of a massive *Deinotherium*, the elephant's shovel-tusked relative, found near the bottom of Bed II on the same site. Both were surrounded by stone tools. Whether these behemoths had been hunted down or scavenged, there seemed little reason to doubt that the hominids of Olduvai Gorge were getting their hands on large quantities of meat. Apparently, *Homo habilis* had already embarked on a hunting way of life, almost two million years ago. Just a little farther up in the Olduvai stratigraphy came evidence of even bolder ancestors: a skullcap of *Homo erectus*, and the fine Acheulean tools associated with an increasing frequency of larger mammal remains.

Louis Leakey drifted away from work in Olduvai during the 1960s, leaving Mary and a steady stream of visiting archeologists to elaborate Olduvai's crucial contributions to the "hunting hypothesis." At the same time, Clark Howell was busy excavating two sites in Spain that would provide abundant evidence that man had become a confirmed hunter some 400,000 years ago. At a place called Torralba,

Howell discovered the remains of almost thirty elephants, along with various skeletons of rhinos, red deer, horses, and wild oxen. Nearly eight hundred Acheulean tools punctuated the debris—and even some weapons made of wood. No hominids were found, but based on the time period and the form of the tools, it was safe to assume that the inhabitant of the site was *Homo erectus.* The archeologists also discovered bits of charcoal suggesting the use of fire, and a strange logic to the placement of some of the bones. Almost the complete left side of one elephant skeleton was found arranged as if for display, each bone turned over and replaced in the position it would have held in life. At the nearby site of Ambrona, Howell found several elephant leg bones lying end to end in two perpendicular lines.

Along with Leslie Freeman, Howell pieced together a scenario that neatly accounted for the complexion of the sites. Apparently, bands of hominid hunters had deliberately set fires to the grassland, driving a migrating herd of elephants into a bog. Mired and weakened in the swamp, the elephants were pelted with rocks from above, then finished off with spears. The hunters closed in to carve up the carcasses with stone tools. The perpendicular lines of leg bones found at Ambrona may have been a walking bridge of elephant limbs, set up to haul the bonanza of flesh out of the muddy swamp. The oddly symmetrical half-carcass was harder to explain—perhaps it was the remnant of some ritual, though no other signs that *Homo erectus* indulged in ceremony had ever been found.

Vividly illustrated by painter Stanley Meltszof in Time-Life books on early man, the Torralba-Ambrona massacres epitomized the "Mighty Hunter" vision of our ancestors rapidly gaining strength in the late 1960s. But the hunting hypothesis was much more than a supposition that humans had been hunting for a very long time. It was a unifying theory arguing that hunting was what made us human in the first place.

"Human hunting is made possible by tools, but it is far more than a technique or even a variety of techniques," wrote Sherwood Washburn and C. S. Lancaster in *Man the Hunter,* in 1968. "It is a way of life, and the success of this adaptation . . . has dominated the course of human evolution for hundreds of thousands of years. In a very real sense our intellect, interests, emotions and basic social life—all are evolutionary products of the success of the hunting adaptation."

Adding force to this view were new studies of contemporary hunter-gatherer societies, principally the !Kung San and G/wi people of the Kalahari Desert, the Hadza of Tanzania, and the aborigines of the Australian outback. Unlike earlier observers of these "primitives," anthropologists like Richard Lee and Irven DeVore did not hold these societies to be quaintly preserved remnants of Stone Age Man. They believed, quite rightly, that the hunter-gatherers alive today had developed patterns of existence unique to *Homo sapiens* and, incidentally, just as viable for their purposes as those of other modern men. They reasoned that since the hunter-gatherers did not practice agriculture, however, their habits—especially as they pertained to procuring food—might reveal something about the principles governing the lives of ancient hunters as well.

One principle that seemed to be universal among hunter-gatherers was a distinct division of labor between the sexes. The males did most of the hunting, while the females took care of the bulk of the gathering. This division of labor, completely unknown among the other primates, formed the central post of the hunting hypothesis. A whole constellation of uniquely human behaviors spun out from it. Hunting and butchering the dangerous game found at Olduvai and Torralba would require efficient cooperation among males—something rarely found in nonhuman primates. Unlike their ape relatives, human hunters also had to maintain large territories, which would put a premium on learning and on a rising capacity to store knowledge of the environment. Increasing brain growth, more sophisticated weapon making and communication skills, the sharing of food, the origin of monagamous pair bonding and the nuclear family—everything essentially human could be traced back to hunting. Warfare too, of course—the killer ape lived on underneath it all. Washburn and Lancaster also proposed that in the symmetry of the Acheulean hand ax lay the beginnings of aesthetics—an appreciation of an object as beautiful beyond its function.

"For those who would understand the origin and nature of human behavior," they concluded, "there is no choice but to try to understand 'Man the Hunter.' "

No choice? I thought about that for a minute, as I listened to the silence gathering around me in the Junction and watched my empty stage. I had no trouble conjuring up some ancient hunters who would fill the space and fit the scenario: a band of near-men armed with hand axes and spears, huddling in a circle to plan their strategy, while a dimwitted *Deinotherium* grazed nearby, oblivious to its impending doom. The scene was easy to imagine, of course, because I had seen it illustrated in textbooks and popular magazines so many times that it had established a kind of verisimilitude simply through repetition. I had to admit too that the hunting hypothesis had potent "explanatory power"—a phrase used to describe a theory that ties together a lot of loose ends. Many anthropologists today still believe in its essentials. But *no choice*? I shivered at that. If the last twenty years of discovery have taught us anything, it is not to trust *anything* so faithfully.

In fact, there were plenty of other choices waiting in the wings—and critics gathering in the audience. One aspect of the hypothesis that drew a lot of fire in the early 1970s was the emphasis it blithely conferred on hunting. Female brains are of course as large as male brains: Why assume that human cerebral capacity was spurred on by an essentially male activity? Richard Lee's own studies among the !Kung had also shown that hunting accounts for only some 30 percent of that group's diet—the rest is supplied by the nuts, tubers, and other plant material gathered primarily by females. The same seemed to be true among other modern hunter-gatherers. If hunting was apparently a secondary activity, why grant it such overwhelming significance? In sum, how could any theory explain the evolution of a species when it ignored the contributions of 50 percent of the population?

Feminist anthropologists, notably Nancy Tanner and Adrienne Zihlman, sized up these shortcomings of Man the Hunter and offered in his place a version of our beginnings launched by Woman the Gatherer. They contended that the prime mover for the split from the apes was indeed the development of tools—not weapons for hunting, however, but digging sticks, hammerstones, and various containers, all used to gather plants, eggs, honey, termites, and perhaps an occasional burrowing rodent. In this hypothesis the fundamental social unit was that of mother and child, with the female sharing the fruits of her foraging with her long-dependent offspring. Through generations, this economic alliance between mother and child might expand to include sharing between the gathering female and other kin, and eventually a more generalized sharing with adults outside the kin group, including males. Since the most important food sources were gathered, however, it would have been females, not males, who were in charge of the technical innovations that nurtured the growth of intelligence and set *Homo* apart from the other primates. Hunting played at most a peripheral role: that of a hit-or-miss, inefficient activity that males could afford to pursue because they could count on the women back at the living site to provide the steady source of food.

There is no doubt, in my mind at least, that the gathering hypothesis helped to correct a view of human origins that was woefully out of balance. (Mentally, I shooed my band of hunters offstage. To take their place I ushered in a clutch of resourceful she-hominids. Their young ones watched, engrossed, as they coaxed tubers out of the Olduvai earth.) But there were problems with this theory too, not the least of which was the replacement of one sex bias for another. Why should a female activity be any more exclusively responsible for the development of human intelligence than a male one? Certainly plant foraging played an extremely important part in early hominid evolution—but it doesn't necessarily follow that we have to shrink the role of the Mighty Hunter to that of an Idle Sportsman. It is simply a fact that in all present-day hunter-gatherer societies the males in the social group *do* hunt, while the females by and large do not. If hunting served no purpose, it is very difficult to see how it would have survived so long in evolution, let alone how it might have developed in the first place. The hunters would have long since disappeared, outcompeted by the industrious males who spent their time in more profitable pursuits—like foraging along with the females.

Another shortcoming in the gathering hypothesis is the lamentable lack of evidence either to support or refute it. Meals of meat and marrow leave bones behind; tubers and termites leave nothing. Likewise, a stone tool is forever a stone tool, while a digging stick is very soon a pile of dust. Until we can find some way of more accurately quantifying the amounts of meat and plant matter consumed in the past—and some interesting progress is being made in that direction—the hard archeological evidence of bones and stones will carry more weight than the most elegant speculation.

Which brings us back to the real stuff of FLK. In the late 1970s, the litter on the Zinj floor, along with what had been discovered at some sites at Koobi Fora, provided the props for another version of humankind's beginnings—balanced, persuasive, and enormously appealing. The designer of the new hypothesis was Glynn Isaac, the tireless and influential co-director of Richard Leakey's expeditions for many years. Isaac began by asking the same question that set off these reflections: What was it that our ancestors were doing differently from other primates that led them to become human? He reasoned that "using tools" was not an answer, since it begged the question

The excavation at FLK today. At center is the marble plaque marking the spot where Zinjanthropus was found in 1959. DONALD JOHANSON

of what the tools were being used *for*. "Hunting," however mightily, was not a satisfactory answer either. For one thing, new field research on chimpanzees and savanna baboons was revealing that among primates, humans were not so unique after all in their taste for flesh. Modern hunter-gathers clearly hunted more systematically than nonhuman primates, and with larger animals for prey, so hunting might be part of an answer. But for Isaac, the hunting hypothesis put the emphasis in the wrong place.

In an outcrop about fifteen miles east of Lake Turkana, Isaac and his co-workers had uncovered the remains of a hippopotamus that had died in a delta channel some 1.6 million years ago. Among the hippo's bones were over a hundred stone tools, mostly small flakes. Several cores were found, along with a battered river pebble that evidently had been used to knock the flakes off the cores. Whether the hominids had killed the hippo themselves or merely come across its dead carcass was not really important. (Given the primitive forms of the tools, Isaac suspected scavenging.) What mattered was that no other stones in the sediment were larger than a pea, so it was certain that the raw material for the tools had been *carried* from somewhere else, then fashioned into usable implements on the spot. At another site nearby, Isaac's team found more stone tools, together with a scattering of mammal bones—hippo, giraffe, pig, porcupine, and various bovids. Unless one presumed that the toolmaking hominids had simply happened upon a menagerie of dead or dying animals, then this site had to represent a living floor—or "home base," as Isaac put it—and so did the Zinj level at FLK. The bones had been *transported* here, presumably when they still had meat on them.

For Isaac, that was the crucial issue. In all known human societies, "bringing home the bacon" (not to mention the vegetables) is an utterly commonplace activity. Isaac concluded that the most reasonable explanation for transporting food away from its original location *is to share it with somebody else.* Like the proponents of the hunting hypothesis, he believed that a division of labor was the critical human innovation. But the methods of obtaining food—hunting, scavenging, gathering, whatever—were not as significant as the act of sharing itself. At its most fundamental level, a subsistence strategy based on food sharing would provide the economic advantage of expanding the resources available to all members of a group. This economic security would be especially important to a species whose females were kept closer to the home base by young who stayed dependent for a long time.

The beauty of Isaac's theory was that it did not account just for the home bases evident at FLK and the Koobi Fora sites—it made human evolution impossible without them. In time, the economic advantage of sharing food would give way to an array of uniquely human social behaviors. For instance, hominid groups who could share not only food but *information* on the location of resources would be favored by natural selection over those who could not—leading to the development of language. Food sharing might also play an intrinsic part in the development of a sense of reciprocal obligation, giving way to the extremely complex social and economic alliances found in human societies. Finally, the food-sharing hypothesis could also help explain human marriage patterns, as an outgrowth of the division of labor that allowed males and females more efficiently to exploit the available resources in their environment.

We're ready now for another change of set on my ancient proscenium. I did not have to use my imagination to bring a food-sharing scenario into being; Isaac wrote one himself in 1976. He meant it to describe life at Koobi Fora in the early Pleistocene, but it will play just as well on the FLK stage. The scene is the swampy shores of a lake, full of hippos, crocodiles, and birds. Beyond stretch some open floodplains:

Far across the plains, a group of four or five men approach ". . . *striding along, fully upright, and in their hands they carry staves.* . . . *As the men approach the observer becomes aware of other primates below him. A group of creatures have been reclining on the sand in the shade of a tree while some youngsters play around them. . . . They seem to be female, and they whoop excitedly as some of the young run out to meet the arriving party . . . the object being carried is the carcass of an impala and the group congregates around this in high excitement; there is some pushing and shoving and flashes of temper and threat. Then one of the largest males takes two objects from a heap at the foot of the tree. There are sharp clacking sounds as he squats and bangs these together repeatedly. . . . When there is a small scatter of flakes on the ground at his feet, the stone worker drops the two chunks, sorts through the fragments and selects two or three pieces. Turning back to the carcass the leading male starts to make incisions . . . each adult male finishes up with a segment of the carcass, and withdraws to a corner of the clearing, with one or two females and juveniles congregating around him. They sit chewing and cutting the meat with morsels changing hands at intervals. Two adolescent males sit at the periphery with a part of the intestines. They squeeze out the dung and chew at the entrails. One of the males gets up, stretches his arms, scratches under his armpits and then sits down. He leans against the tree, gives a loud belch and pats his belly.*

For all their bad manners, Isaac's hominids were more generous and humane than the mighty hunters who had preceded them on stage and infinitely more appealing as ancestors than Dart's demon apes. The food-sharing hypothesis, embraced by Isaac's friend and colleague Richard Leakey in a series of popular books, enjoyed tremendous acceptance in the late 1970s, and still does in some places. Is it just a coincidence that the late 1970s was a time when modern males too were beginning to embrace a more sensitive, caring self-image? I don't think so. Nevertheless, no one before Isaac had so neatly accounted for the hard evidence of bones and stones.

Though perhaps not neatly enough. Sitting on that mound of excavated earth by FLK, I decided to add a little coda to Isaac's published scenario. I pictured again those hominids hunched over their shared meat and entrails. Unbeknownst to them, another figure was approaching across the floodplain, also walking upright, with just a hint of a swagger in its gait. As the figure came closer, I could see that it was wearing baggy workman's trousers and a lightweight sweater. A dirty blue bandanna was tied around its forehead. The figure stopped in front of me and leveled a cold stare at the happy hominid group. Then it turned to face me, shaking its head.

"You can believe this poppycock, if it makes you feel all warm and gooey," said the figure. "But if you want to explain human evolution, you got to *go for it.*"

Lewis R. Binford had arrived at FLK, and the stage would never be the same again.

Throughout his career, Lew Binford has never once been praised for his humility. As a grad student at the University of Michigan, his contempt for current fashions in archeology so alienated his thesis adviser that the man declared Binford would receive his Ph.D. "over my dead body." Binford got the degree. From Michigan he took a teaching position at the University of Chicago. His first lecture there nearly caused a riot; he was not allowed to finish. Later, he team-taught a course with the eminent Robert Braidwood, a doyen of Middle Eastern archeology and a powerful force in his department. The forty-three students in the course were given the choice of writing papers for Braidwood or Binford. Forty-one chose Binford.

Lew describes himself in those early years as "intolerably arrogant." But sometimes arrogance is just what is needed most to shake up a science. His generation of archeologists had inherited a very idealistic approach to understanding the past. Their more traditional predecessors had long held to a view that placed enormous value on the role *culture* plays in determining the material record from the past. The artifacts that archeologists deal with—stone tools, potsherds, broken bones, and the like—were looked on as cultural texts, left behind by primitive people almost in the same way that later individuals left behind documents, diaries, and other accounts that form the historical record. But whereas historical texts could be read and interpreted, the leavings of prehistoric people were essentially mute because there was no way of ever knowing how the minds that produced ancient cultures were working in the first place. The best that could be done was to examine the *forms* of tools and other artifacts and organize them into categories that would in essence reveal the cultural progress of the human species through time—leading (naturally) to the pinnacle achieved by modern Western civilization.

Early on, Binford decided he was not about to spend his career poring over the forms of artifacts and assigning dates to collections. He and some other young investigators suspected that some new approaches to site interpretation could squeeze more meaning out of the stones and potsherds than their mentors realized. They began to look for new ways to see, searching for patterns in the distribution of artifacts on a site, for instance, or analyzing stone tools for clues to their manufacturing techniques and original function. But what was needed was more than a few new tricks in a researcher's bag of techniques: Archeology required a complete new set of rules and a rejuvenated sense of itself as a science.

One of the notions dispelled by Binford's "New Archeology" was the false assumption that the job of archeologists was to "dig up the past." If you think about it, no shovel, pickax or bulldozer is capable of digging up the past, unless the implement has a time-machine attachment. Archeological sites exist *in the present;* they are as much a part of the contemporary world as the rocks, trees, houses, and billboards that also occupy a given landscape. Furthermore, an archeological site is static and lifeless. On its own, it cannot say anything meaningful about the living processes that brought it into being. In Binford's view, archeologists had been consistently confusing the past and the present when they came to the task of interpreting a site—a messy situation made all the messier because they failed to realize that the observations they were making were *also* part of the contemporary world.

Let me give you an example of where this might lead. In their summation of the hunting hypothesis, Sherwood Washburn and C. S. Lancaster had speculated that the beautiful symmetry in Acheulean stone tools may have developed because symmetrical projectiles when attached to a shaft, travel more accurately through the air than lopsided ones. Likewise, a hand ax fixed on the end of some kind of handle is easier to balance when it is equally weighted on both sides, so symmetry has a functional advantage there too. This observation is undoubtedly true and makes exquisite good sense. But it is nonetheless an observation made in a *contemporary* landscape, one that, to put it crudely, includes not only an ancient hand ax site, but a modern hardware store too, where a selection of good-quality ax handles can be found in aisle three. Considered on its own, there is nothing about the symmetry of an ancient hand ax that suggests a handle was once attached to it, or that the implement was used to clobber antelopes, skin elephants, or mash potatoes. Come to think of it, there is nothing about the symmetry of a hand ax that says anything at all, beyond the obvious statement that its maker was capable of fashioning a symmetrical tool.

In Binford's world, what you know about such things is much less important than what you don't know. Put another way, the first, crucial task of any scientist is to deliberately and unsparingly take stock of his own ignorance—because only when he knows the level of his ignorance can he begin to carve away at it, leaving the truth behind like a sculpture brought forth from raw stone. Instead of wasting time building theories on what we have already assumed to be true, we should instead cast a critical eye on those assumptions and see if they hold up to the facts.

Binford accused his peers of making one common assumption in particular: that everything found associated with human artifacts is by definition a product of human activity. Remember that pile of dik-dik feces back at the new hominid site? Let's say that a similar pile became fossilized a couple of million years ago. (Fossilized feces are called "coprolites," and they are by no means rare.) Now let's say that in the process of excavating a living floor at Olduvai Gorge, an archeologist uncovered that long-buried pile. He finds it surrounded on all sides with stone tools and flakes, and nearby is a scattering of broken bones. If he knows anything about dik-dik behavior, he will quickly recognize that the concentration of feces was a defecation spot for an ancient pair of dik-diks. *But if he is ignorant of the natural process that forms such piles,* he might well assume that the hominids who once lived on the site had collected the pellets, one by one. Based on that commonsense inference, the archeologist might then begin to build a plausible argument that Man in the early Pleistocene had already developed the slingshot, and had tapped into a naturally available supply of ammo.

Nobody would be that silly, of course. But as Binford pointed out, plenty of people were assuming that the existence of a bunch of broken bones together with a bunch of stone tools at such sites as FLK meant that hominids had used the tools to break up the bones. In fact, any number of natural, nonhuman forces might account for bone concentrations. Perhaps they were carried there by the waters of a stream. Perhaps, like the bones of Makapansgat, they had been accumulated over time by carnivores. Before they were buried, the bones could have been broken up by passing herds of animals, or they might have been crushed after burial by geological forces. Perhaps hominids were responsible for *some* of the broken-up bones, but not all. How could an archeologist discriminate which ones? How could he accurately reconstruct the *dynamic* processes of the past, based on the lamentably *static* record that survives in the present?

Binford's answer to that question is called "middle-range research," a term he invented to embrace any kind of investigation needed to justify whether an assumption about the past is true or false. Most middle-range research involves looking at the *present.* If this seems like an odd way for archeologists to be spending their time, keep in mind that living human beings and animals, like their counterparts in the past, leave behind an archeological record too—but their "static" litter can be traced back to real, observable, dynamic behaviors. Once something quantifiable is known about the relationship of modern processes to the litter they leave behind, then you can feel more secure about reconstructing ancient processes from the remnants they left behind too. Take for instance Bob Brain's research on the bone accumulations in the South African caves—a classic example of middle range research. Instead of assuming that the bones had been accumulated by ancient hominids, Brain looked to the modern world for living processes which would create similar bone concentrations. In that case, it turned out to be leopards feeding in trees.

While Brain was digging into the mystery of the South African caves, Binford was applying middle-range research to another enigma, trying to make sense of the puzzling variations in the stone tool kits of Neandertal Man in Europe. He studied the bone rubbish of Navaho Indians and the debris left behind by stone knappers in Australia. His quest sent him to the Arctic, where he spent years with Nunamuit hunters, examining the cut marks, patterns of bone fractures, and the distribution of bones left behind when they butchered a kill. At the same time, he began to accumulate data on wolf kills, hyena dens—any activity, human or otherwise, that might influence the complexion of litter on a modern landscape. Somewhere along the way, Binford came to realize that the huge amounts of modern-day data he was accumulating might have some bearing on FLK and the other early Pleistocene sites. It was hardly what the specialists wanted to hear.

"Let's face it, you guys were getting away with murder," Binford told me once before he left the Gorge. We were sitting around the camp table in the dark, smoking some small cigars I'd brought along. "Mary Leakey, Glynn Isaac—the whole African group didn't even know they had a problem. They'd just go out and have a good time, and when they'd find something, they'd call a press conference and open up the champagne."

Binford's first target was an archeological site in Kenya called Olorgasailie. Glynn Isaac had found seemingly solid proof there of large-scale hunting by *Homo erectus* some 400,000 to 700,000 years ago. His most dramatic evidence was an area less than fifty feet in diameter, where the shattered remains of ninety giant baboons lay together, along with a large number of hand axes. Isaac's interpretation: The baboons—an extinct species the size of female gorillas—had been slaughtered and butchered en masse by a band of hominids.

"I jumped all over Glynn for that," Binford said. "There was no evidence at all that those baboons had been killed all at once, or even that they'd been hunted by hominids at all." He showed that the pattern of bones could have been influenced as much by running water as by hominid hunters. Other smashed-up bones Isaac had cited as evidence of hominid predation might just as well have been broken up by trampling hooves or other nonhuman forces.

Having drawn blood at Olorgasailie, Binford charged headlong into battle against Man the Mighty Hunter. The oldest *indisputable* evidence for systematic hunting by hominids was Clark Howell's famous elephant massacre site at Torralba; Binford disputed it. He agreed that there was a pronounced concentration of large animal remains on the site, especially elephants. But even Howell admitted that the deposits in which they were found could have accumulated over a great length of time. Since the area was a swamp, and elephants are known to migrate to marshy areas to die, why rule out the possibility that these animals might have died natural, insignificant deaths over the centuries, unembellished by human intervention?

"Clark had observed that these elephant remains were found along with lots of Acheulean stone tools," Lew told me, helping himself to another cigar. "So he and Les Freeman developed that dandy four-color picture of mighty hunters butchering herds of elephants. Les even calculated the amount of meat on that many elephants and estimated that it would have taken fifty men six trips apiece to cart the meat out. So now you've got *Homo erectus* organized into cooperative labor groups, half a million years ago! Well, I showed that the tools they found at the site weren't really associated with the elephant bones. In fact, there's an inverse correlation—where you find the elephants, you don't find the tools. The artifacts were actually correlated with the lower limbs and mandibles of red deer, bovids, and horses—what you'd expect to find if the hominids were scavenging, not hunting. So much for the classic mighty hunter site."

Armed with his heavy core of data—honed to a fine cutting edge by sophisticated statistical techniques—Binford hacked away at the evidence for hunting at sites covering four continents and extending from the early Pleistocene all the way to the upper Paleolithic, only some 30,000 years ago. All over the world, he believed archeologists had grossly overestimated the sophistication of the ancient people they were seeking to understand.

But he saved his most pointed attack for Olduvai Gorge—which brings us back to our FLK theater. You'll remember we had Glynn Isaac's hominid family up onstage, with Lew Binford standing to one side, giving them the cold eye. Under his skeptical gaze I saw the apparitions quiver and disappear, leaving Lew alone, hands stuffed in his pockets, probably puffing on one of my cigars. I didn't have any trouble imagining this, because Lew and I had visited FLK together a day or two before the discovery at Dik-dik Hill put an end to such sightseeing. He had his own Pleisto-drama ready to replace Isaac's, executed with the sweeping bravado of a Broadway veteran come to show the yokels how to act.

"Okay, Don, we're positioned now on the Zinj occupation site, alias Mary's living floor, alias Glynn's home base," he said, as if he were briefing me on an impending military encounter. "How do we know it's a home base? Because it's got lots of tools and broken bones, right next to where a lake used to be. But think about it, Don—why would these hominids want to establish their camp next to a lake?"

"There's bound to be a lot of other life around," I answered, "and obviously a water source . . ."

"That makes sense for Boy Scouts in Yosemite Park, but if you're some three-foot diurnal biped out on the Pleistocene savanna, it'd be the worst place to be. The lakeshore is where the lions and other predators will be hanging out, waiting to eat you. Even today, hunter-gatherers in remote regions, with fire and bows and arrows and other sophisticated weaponry, almost never camp by a water source."

"Maybe they were sleeping somewhere else," I countered. "They could still be bringing food back to a central place to eat."

"Sure they could. And if you just want to feel all warm and cozy, you could even imagine them sharing the food around in the social group. The alternative is to try to justify those inferences. Get your hands dirty in some real science."

I'd read Lew's book, so I knew what he was getting at. If we take as the inference to be tested the existence of a home base where hominids were transporting meat to be eaten, then that dynamic process should have left some very specific trace of itself in the static archeological record. For instance, among the debris on the site there should be more meat-bearing bones, like the upper limbs of bovids, than the total number of other kinds of bones can account for. Before you can make that judgment, however, you first have to factor out all the *other* dynamic processes that added their traces before the site was buried. Hyenas may have carried some bones away, or brought others in, or gnawed the ends off still others. Perhaps some carnivore made its den here once. And there's always stream action to consider. Once you've subtracted the contributions of all the nonhominid agents from the Zinj-floor bone pattern, the bias in the representation of one kind of bone over another can be attributed to the actions of hominids with some real confidence. It's not just an assumption anymore.

"I went to work on Mary's data, and sure enough, there was a surplus of some anatomical parts among the bones on the Zinj floor," Lew said. "But the surplus wasn't in the meaty upper limbs predicted by the home-base model. Quite the opposite. There were significantly more *lower* limb bones, as well as skulls and mandibles, than could be accounted for. The hominids were transporting carcass parts, all right. Only they were carrying around the pieces of a skeleton that would yield the least amount of meat!"

Standing on the spot where it had all happened two million years ago, Lew then offered his own version of the FLK play.

"In my opinion these guys were continuously feeding, moving through the landscape like a damn troop of baboons. Tubers, grubs, seeds, whatever. So let's say one of them looks up and sees some vultures circling. With a few whacks on a core he makes one of these choppers, then off he goes. He gets to the kill site, but by this time there's been lots of scavengers and there isn't much left of the prey. All that's left of any value is the little bit of marrow in some lower-limb bones, but to get at it he has to soak the limb in water to soften the skin. So he carries the thing over to the lake shore. Then he smashes up the bone, gets what he wants, and leaves tools, bones, and everything else behind. That's the stuff we find at FLK. There's nothing suggesting a living floor, a home base, a thorn windbreak, a social system based on sharing, or anything other than a place where some tool-using creatures broke open some already defleshed bones to dig out the morsel of marrow inside."

Naturally, Binford's trashing of the conventional wisdom was not based on his interpretation of the Zinj level alone. He also de-

Lewis Binford holds forth at Olduvai Gorge.
DONALD JOHANSON

bunked Mary Leakey's elephant and *Deinotherium* sites nearby, the famous DK stone dwelling, Isaac's hippo site near Lake Turkana, and anything else recovered from *Homo habilis*'s time on earth that vaguely suggested large-scale meat-eating. In Binford's view, there was indeed a battle for meat going on here in the early Pleistocene—but the combatants were well-established, specialized predators like lions, saber-toothed cats, and hyenas, not hominids. After all the other meat-eaters had had their fill of a carcass, the hominids scuttled in last and made off with what little scraps of marrow remained.

"It's clear from the bone distributions and the way the limbs had been busted up," Lew told me on the site. "The Olduvai toolmaker was no mighty hunter of beasts. He was the most marginal of scavengers."

So much for romance.

Lew Binford is endowed with infectious conviction. By the time we walked away from FLK that afternoon, I had become a believer: *Homo habilis* was an unspecialized opportunist whose tool-assisted scrounging of pre-ravaged carcasses had little to do with "becoming human." But now Lew was gone, and except for an crowned plover darting in and out among the gullies, the old excavation was empty. The scene was set for the final act of the drama. Before it was over, my FLK stage would start to look more like a boxing ring.

While Binford was preparing his full-scale assault on Olduvai Gorge, Glynn Isaac was hardly waiting around for the ax to fall and crush his food-sharing tableau. He had a knot of talented students working full tilt, principally at a new site near Koobi Fora. Called FxJj50—or Site 50 for short—the new excavation had yielded more than fifteen hundred pieces of worked stone and two thousand fossilized bone fragments strewn about a river bank 1.5 million years old. What was intriguing about the site were the indications that it had been occupied by hominids for only a very short time, then quickly and gently buried. In this regard, it provided an even better "snapshot" of ancestral life than the jumbled and overabundant record at FLK.

Working Site 50 with Isaac was a young man named Nick Toth, who later joined us at the Institute and is now on the faculty at Indiana University. Toth's target of research was the collection of stone artifacts, in particular, the patterns of fracture evident on their surfaces. To make sense of their meaning, Nick taught himself to make stone tools and used the results of his handiwork to butcher animal carcasses. This sort of modern-day modeling of hominid tool use was nothing new in itself. (Louis Leakey, for one, was wowing Olduvai visitors with his stone-age butchering techniques before Nick Toth was born.) But the conclusions were radical.

Most archeologists had been focusing on the various forms of the worked tools—the discoids, end choppers, side choppers, and so on laid forth in Mary Leakey's descriptions of the Oldowan artifacts. What Toth found was that these forms were simply the result of knocking small, sharp flakes off a core. The little flakes themselves, which previous investigators had been interpreting as the waste by-product of tool manufacture, were in fact fairly handy implements for cutting up a carcass. It Toth was right, *Homo habilis* would not have needed a great deal of stone-knapping expertise to gain access to the meat beneath an animal hide.

Just because the flakes *could* cut meat did not mean, of course, that they were being used for that purpose. While Nick was busy at Site 50, Larry Keeley from the University of Illinois at Chicago Circle was examining under a microscope some flint tools from upper Paleolithic sites in Europe. Keeley had found traces of "micropolish" on the cutting edges of the tools, indicating that some had been used to cut meat, while others had been variously applied to bone, hide, wood, or some softer plant material. Toth wondered whether the same kind of discriminations could be found on the much older Koobi Fora tools. The two archeologists decided to collaborate.

Unfortunately, most of the Koobi Fora tools had been made of rough lava, a stone too coarse and too sensitive to weathering to show signs of micropolish. But Toth was able to find fifty-nine finer-grained tools in the collection, mostly flakes of chert. Keeley's microscope revealed micro–wear polish on nine of them—four with polishing marks indicating animal butchering, three characteristic of woodworking, and two others showing the sort of polish produced by contact with soft vegetation.

Out of the hundreds of thousands of stone artifacts from the early Pleistocene, the testimony of one technique focused on nine little flakes was hardly enough to inspire a chorus of "eureka,"even among the food-sharing faithful. Nevertheless, it provided a tiny morsel of evidence for meat processing, much more useful in its way than the beautifully reasoned "Just So" stories abhorred by Binford. Meanwhile, another Isaac protégé, Henry Bunn, now at the University of Wisconsin, was squinting over the bones collected from the site, looking for the telltale cut marks and fracture patterns that the hominids' tools would have left behind. His research eventually led him to the FLK bones in the Kenya National Museum in Nairobi—and smack into the path of Binford's *hominid* scavenger hypothesis.

Unless a bone is very quickly buried, it will bear on its surface the markings of its fate after death. The scars on a fossil bone fall into three general categories: cut marks and fractures left by hominids wielding stone tools; gnaw marks and tooth punctures recording the munchings of carnivores; and the pits, cracks, and scratches produced by natural forces such as weathering and abrasion. People had been interpreting these surface hieroglyphics in paleo-Indian remains for years, but until Bunn and others came along a decade ago, nobody had paid much attention to them in older collections.

Bunn had little trouble distinguishing the cut marks on the East African bones from the gnaw marks that were also present. Bob Brain's work in South Africa had shown that a stone tool leaves a distinct V-shaped groove in the bone, while a carnivore's tooth leaves a broader, U-shaped impression. Not surprisingly, Bunn found plenty of both kinds of damage. What was interesting was the *location* of the tool-inflicted cut marks. In many instances the cut marks were clustered around the articular ends of upper-limb bones—precisely where a hominid with a tool would scrape against the bone when attempting to detach meaty limbs from a carcass. Some of the upper limbs were also severely bashed up, enough to convince Bunn that the bashers were almost certainly hominids wielding hammerstones, going after the big bones' rich trove of marrow. If the hominids were butchering substantial amounts of meat at particular places on the landscape, then whether or not the meat had been hunted or scavenged, Isaac's food-sharing hypothesis looked pretty good after all. So much for marginal scavengers. In his report in *Nature,* Bunn was icily exact about the implications of his work for Binford's theory: "This direct evidence of early hominid diet," he wrote, "allows us to dismiss models of human evolution which do not incorporate meat-eating as a significant component of early hominid behavior."

Henry Bunn was not the only young investigator burning the midnight oil in the museum in Nairobi. At the same time, Rick Potts of Harvard University and Pat Shipman of Johns Hopkins were also taking a closer look at the cut marks on the Olduvai bones. Using a scanning electron microscope, Potts and Shipman discovered that different kinds of tool use were distinguishable by microscopic features of the mark left behind. Tools used for slicing, for instance, left fine, parallel grooves within the V-shaped cut mark. Like Bunn, Potts and Shipman believed their evidence proved that the hominid tools and the faunal bones at sites like FLK weren't there together by accident. But they disagreed with Bunn on some key observations. Whereas Bunn had looked at the thirty-five hundred FLK bones and found cut marks on more than three hundred of them, Potts and Shipman could identify only a few dozen.

Most important, they also disagreed on the critical issue of where those cut marks were located on the bones. Bunn said most were on the articular ends of meaty limb bones, but Potts and Shipman found a greater number on the bone shafts—an odd place to be slicing away with a tool if your aim is to detach flesh from a carcass. Furthermore, Potts and Shipman could find no indication that the hominids had given any more attention to meaty bones than they had to the meat-poor lower limbs. They concluded, contrary to Bunn, that there was no evidence at all that hominids at Olduvai and Koobi Fora had been systematically butchering meals of meat. Bunn countered that they had simply been counting the wrong things. Trained as paleontologists, Potts and Shipman had included in their study only bones whose taxon could be identified. Bunn himself, on the other hand, was using every scrap of bone on the site in his calculations, nameable or not.

When it came time to expand upon the meaning of FLK and the other Olduvai sites, not only did Potts and Shipman disagree with Bunn—they disagreed with each other as well. In his Ph.D. thesis Potts analyzed the whole complexion of Bed I archeological sites and came up with a novel alternative to the traditional home-base model. Using a sophisticated computer modeling program, Potts showed that foraging hominids would have saved energy by "caching" stone tools at specific locations throughout their territory.

Whatever food they found could be carried to these caches to be processed. Rather than representing the remains of a home base, the Zinj floor was perhaps the vestiges of just such a stone cache, frequented over and over again through a period of several years.

Pat Shipman's view swung even further away from the popular wisdom. Since her analysis had shown that the cut marks on the Olduvai bones were not where they *should* be if the hominids were stripping meat, then either the hominids were very stupid or they were after something else on the bones besides the meat. Shipman searched through the literature on modern hunter-gatherers and discovered that cut marks encircling lower limb bones are commonly left behind when a carcass is skinned. There were no encircling marks on the ends of the fossilized lower limbs, but there were plenty of cut marks there nevertheless. Shipman concluded that the hominids were scavenging carcasses and removing *tendons,* perhaps to use as thongs in the making of primitive carrying devices. They were probably eating what little meat was left on the bones as well, and perhaps using the skins too to make "sack-like carrying containers."

Shipman had also noticed that among the Olduvai bones were a few with hominid cut marks laid directly *over* the toothmarks of carnivores—direct evidence that these little ancestors were scavenging the kills made by others. Gradually a more elaborate portrait of "Man the Scavenger" began to emerge in her mind: Unlike hunters, scavengers need a very efficient means of locomotion over long distances, so they can search for carcasses. Walking upright would provide just such an energetically efficient means of getting about, though it is a pitifully poor adaptation for the short bursts of speed needed in hunting. Bipedalism would also leave the hands free to carry the tools needed to exploit whatever carcasses one might find. Finally, a confirmed scavenger would need to have some other source of food to fall back on in those trying times when carcasses were hard to come by. Pat Shipman's husband, Alan Walker, had already done electron microscopic studies of fossil hominid teeth, which he believed suggested they ate a diet composed primarily of fruit.

Nowhere in Shipman's view is there room for a home base, food sharing or large-scale butchering of carcasses. You might think that Lew Binford, who had been stressing scavenging for years, might be pleased to have found an ally. Think again.

"Shipman's hypothesis is shot through with unjustified inferences and a complete disregard for logic," Lew told me when we were at Olduvai. "She tells us that early hominids were scavengers, but what are they scavenging? Tendons! She argues for a whole shift in a way of life based on extracting tendons to make tote bags. It's nonsense."

"Well, what about Potts's stone cache theory?" I asked.

"Calling it a theory is kind," Lew answered. "What Potts did was invent a computer model that predicted that it would have been more efficient for hominids to cache stone tools around the landscape. And since the computer said it was efficient, it followed that the hominids were actually *doing* it. But there's no evidence at all! It's just a computer's idea of how to live in the Pleistocene."

Binford believes these investigators, like their forerunners, make the mistake of beginning by endowing early hominids with considerable humanity, instead of framing arguments that will help us find out how human our ancestors really were. Stockpiling stone tools for the future and using tendons to make carrying baskets are activities that imply more foresight—or what Binford calls "planning depth"—than we have any reason to believe *Homo habilis* possessed. The same goes for *habilis*'s descendants; wherever Binford casts his eye on the archeological record—at *Homo erectus* sites in Spain and China, or the leavings of premodern *Homo sapiens* in South Africa—he sees a much less capable, less human ancestor at work than we have been led to believe. In Binford's final analysis, the adaptation to hunting, far from forging humanity for millions of years, did not play a significant part in our evolution until a mere seventy thousand years ago, when fully modern humans began to cope with the harsh conditions of Europe by systematically hunting large game.

And what about Henry Bunn, the Isaac student who had "dismissed" his model of human evolution? Initially, Binford had regarded Bunn's critique much the way a six-hundred-pound saber-toothed cat might look upon a three-foot *Homo* juvenile brandishing a pebble. But Bunn was not going to be easily intimidated. Chosen to review the *Bones* book for *Science*—apparently at the suggestion of Glynn Isaac—Bunn praised the book's opening sections on methodology, but pummeled Binford's view on FLK.

Chief among the weaknesses he perceived was Binford's reliance on Mary Leakey's data to draw conclusions about which kinds of limb bones were more highly represented at key sites like the Zinj floor—the meaty upper limbs, or the nonmeaty lower limbs. Bunn pointed out that Leakey herself had stressed that her tabulations should be looked on as preliminary. Binford had disregarded her caution, and by applying "numerical acrobatics" to the incomplete data, had

The late Glynn Isaac at Koobi Fora.
DONALD JOHANSON

come up with a completely unwarranted conclusion—that Isaac's food-sharing hypothesis was dead. In contrast, Bunn himself had gotten *his* Olduvai data straight from the equid's mouth, so to speak, lying on the shelves of the Kenya National Museum. And *he* believed that food sharing was very much alive.

His usual combativeness exacerbated by the perceived insult that a graduate student had been deputized to review his book, Binford fired off a response to Bunn. In the meantime, Isaac himself had begun to back off from the more romantic assumptions of his hypothesis. Most important, whether hominids were hunting, scavenging, or both, Isaac now acknowledged that there was no evidence that they were sharing the proceeds among themselves. If they weren't sharing food, then the concept of a home base was meaningless, and the whole pattern of human social organization he had outlined came tumbling down. He still firmly believed, on the other hand, that FLK and Site 50 proved that hominids were transporting bones and butchering them for meat—on that crucial point Binford was "very probably wrong." Isaac concluded that much more research was needed before we would really know what had happened at FLK.

Tragically, Glynn Isaac did not live to continue that research. In 1985 he died suddenly at the age of forty-seven. With Isaac gone, the mantle of battling Binford fell upon his former students—and in particular Henry Bunn. Together with his wife, Ellen Kroll—another former Isaac protégée—Bunn went back to the bones, employing exactly the sort of middle-range research methods Binford had marshaled to attack the food-sharing hypothesis in the first place. Bunn and Kroll's results, detailing their view of what happened at FLK, would be published in *Current Anthropology* soon after I returned from Africa.

Judging from the proportion of stone flakes that had been retouched, the great number and location of cut marks, and especially the preponderance of meaty limbs on the site, hominid life 1.8 million years ago clearly included the transport, systematic butchery, and sharing of meat. Whether the meat was hunted or scavenged from fierce predators was not really the issue: Both involved "coordinated group activity" and "repeated participation in dangerous subsistence pursuits." Brave *Homo* was back, this time armed with some hard data. Binford's claim that the Zinj site was dominated by meat-poor bones such as lower limbs "should be regarded as a myth that is flatly contradicted by the archeological facts." Binford quickly responded.

"This study has no intellectual anchor beyond an imagined picture of early hominid life," he wrote, later referring to Bunn's evidence as "wishbones."

Nevertheless, he could no longer regard Bunn's studies as a grad student's rearguard attempt to blow life into a dying theory—especially since he sensed that other people were beginning to take them quite seriously. In his reply in *Current Anthropology*, Binford attacked Bunn and Kroll's cut mark evidence, leaning heavily on Pat Shipman's observations that the meaty limb bones at FLK were in fact more often inflicted with the gnawing marks of carnivores than with cuts from hominid tools. He called Bunn and Kroll's conclusions about hominid access to meat-rich limbs "nonsense," unless they could explain how the bones got to the site in the first place. In order to do that, they would have to account for these hundreds of gnaw marks on the FLK bones—an issue that Bunn and Kroll did not address. Previously, Bunn had declared that the simplest explanation for the gnaw marks was that scavenging carnivores had come upon the bones *after* the hominids were through with them.

"That may be the simplest assumption," Binford had countered, "but it's also the assumption that saves the consensus view of the past." Now he pointed out that carnivores rarely gnaw bones where they find them. Instead, they take them off to some protected spot, where they can eat undisturbed. If hyenas were scavenging *after* the hominids were finished with the bones at FLK, then you wouldn't expect to find many gnaw marks, since any bone juicy enough to attract a hyena would have been carried away.

"Hyena behavior is not such a simple thing, and Binford knows that," Bunn says in response. "He is only using the gnaw mark argument to distract attention from the real issue, which is the over-representation of meat-bearing upper limbs on those ancient sites.

Binford is just plain wrong. He just won't admit it. But that's nothing new for him."

Bunn has a right to his confidence in how many upper limbs he sees among the FLK bones—after all, he counted them. Or at least he counted all the bone splinters and used them to determine a reliable estimate of how many whole bones had originally been on the site. It probably won't surprise you that Lew Binford doesn't buy Bunn's method of counting bones. He believes that Bunn, ignorant of the number of splinters formed when a bone is smashed up, has given too much weight to each splinter, and thus overinflated his estimates. Binford has his own way of determining the number of each anatomical part that is represented at a site—a complex analysis taking as its starting point the number of articular ends of bones left on the site, rather than the number of bone splinters. It won't surprise you either that after his analysis is completed and his disc drive has cooled down, his figures wind up supporting the "marginal scavenger" hypothesis.

Binford's densely detailed reply to Bunn and Kroll, published in *Current Anthropology* in 1988, was followed by a longer counter reply from Bunn and Kroll, which was followed in turn by a short note from the editor declaring an "end of the Binford/Bunn-Kroll exchange in these pages." Which is the equivalent of the referee in a fight suddenly throwing up his *own* hands and crying, "*No más!*" And there it stands.

I began this little drama by asking big questions. What made us human? What was going on two million years ago that changed us, rendered us unutterably and forever different from the other animals? We have never suffered from a shortage of answers. Killer ape, mighty hunter, family man—and now, perhaps, lowly scavenger. In a span of three decades, the image of our original selves reflected in the archeological record has shown all the permanence of a view through a kaleidoscope. The hard data of bone and stone have not changed, but like the colored glass shards, they shift and tumble into new mosaics of meaning, borrowing their order from the stress and bias in the theories that guide our sight.

In retrospect, there's no question that the political and social times we have lived through have distorted our view of the ancient past—the harsh realizations of the human capacity for evil that produced the killer ape in the fifties, the imperial, space-exploring *machismo* of the sixties' mighty hunter, so quickly superseded by a vignette of human beginnings as soft around the edges as a John Denver lyric. Does that mean that prehistorians are concerned more with illusion than truth? Is paleoanthropology really just a sort of high theater? I don't think so. The real progress—the chipping away at our ignorance—goes on strangely independent of the theory making, though it seems to depend on the theory makers to bring it into being. Today everybody knows that Raymond Dart's visions of bone-wielding australopithecine cannibals was farfetched; but it was Dart who first had the insight to look for traces of human behavior in a collection of mammal bones. Sherwood Washburn, Irven DeVore, and other promoters of the hunting hypothesis overplayed the role of hunting as the prime mover of humanization, but in so doing, they taught everyone else how to synthesize the evidence from the archeological record, nonhuman primate studies, and research on modern hunter-gatherers.

What Lew Binford has taught us is to beware of our tendency to invest every chipped stone or pile of bone with a human dimension, a human depth. Personally, I think that his marginal-scavenger theory is also extreme. I believe that the Olduvai archeology really *does* tell us something about bone transport and butchering going on at specific hominid activity places. I'm not ready to take it the next step and buy in to a food-sharing scenario—but *something* was going on in the diet of *Homo habilis* that would soon explode into truly human behavior patterns. That explosion—and what our new Olduvai hominid can tell us about it—is what the next chapter is all about. My point now is that whether Binford is right or wrong, we can only progress further toward the truth by following the difficult path laid down by his methods. We can no longer be satisfied with answers that are merely "satisfying."

"The question 'Is this true?' doesn't lead anywhere," Lew told me once. "The question to ask is 'Does this open up new learning opportunities?' "

One certain truth about FLK is that it is not a healthy place to be alone after dark. As I'd been sitting there, the dusk had moved in and swallowed my stage, pulling a cold wind behind it that bit through my light cotton shirt. In twenty minutes it would be pitch dark. The camp staff had lately reported seeing a small pride of lions in the Junction at night—perhaps the same pride that Bob Walter had run across down by the second Fault. I did not intend to wait around to make a positive ID. Besides, it was dinnertime back at the camp, and I was getting hungry.

CHAPTER NINE:

Lucy's Children

Nothing of him that doth fade
But doth suffer a sea-change
Into something rich and strange
—*The Tempest,* I ii

"EAT PLENTY OF BREAKFAST," TIM said. "You'll need the strength. We're going to dig a trench."

We were sitting around the table just after dawn on the last working day of the season. Most of us were only half awake, but for some reason Tim was all fired up, acting like we'd just arrived in the field.

Bob Walter looked up from his coffee.

"A geological trench? I've already dug two trenches."

"And lovely trenches they are. Mary Leakey herself could not have made straighter walls. Only both of them came out sterile. No bone at all. That's been bugging me ever since. You'd think that there would have been a stratum somewhere above the site that held some bone. It's a long shot, but if we were to dig a third trench directly upslope from Dik-dik Hill, maybe we could pinpoint the stratum where this hominid is coming from."

I glanced at my colleagues around the table. Most of them would have looked about as eager if Tim had asked them to spend the last day in camp spooning out the latrine. Frankly, I wasn't too keen on digging a hole into the side of the Gorge myself. There was already too much to do—specimens to be sorted and packed, paperwork that I'd been putting off for days. I was also hoping that we might still have a shot at getting some aerial photographs of the hominid site. The day before, I had received word that a bush pilot named Hans Schneider was to fly in from Arusha and help us out with his two-seat Cessna C150, a plane light enough to land and take off from Olduvai without any trouble.

But Tim is hard to dissuade once he's got an idea in mind. We threw some pickaxes into the back of the Land-Rover and headed down to the site. If the bush pilot made it in, I thought to myself, I would have no choice but to abandon trench digging to the others—with deep regret, of course. I could then spend the rest of the day flying over the Gorge in the Cessna, occupied with the heavy labor of snapping my camera shutter and changing rolls of film.

You may recall that when Bob Walter had dug his original trenches—the first, big one a dozen yards south of the find, the second directly into the sterile southern side of Dik-dik Hill—he'd been trying to get his bearings on where the site fit into the overall geology of the Gorge. It would have been a tremendous bonus if he had found some more of the hominid *in situ* in a deposit above the site, but that hadn't happened. Tim was pinning his hopes on running into some *in situ* hominid bone with his new trench, which would tell us exactly where—and when, geologically speaking—the hominid was coming from.

I'd like to say we took a few swings with the pickaxes and uncovered that distal femur we'd been looking for. But we weren't quite *that* lucky. At the price of a lot of sweat, the new trench revealed very much the same stratigraphic pattern of tuffs that Bob's first trench had shown—with one important addition. Just beneath the coarse-grained ashfall known as Tuff IC, Bob's ax exposed a small, isolated patch of dark sand. This "sand lens" did not extend even as far as the two sides of the trench, so it was no wonder the first excavation had missed it entirely.

Tim knocked some of the sand into a karai with his hammer, poked at it with his finger, and pulled out a fossil. It wasn't hominid, but it was stained the exact brownish-black color as the bones of our hominid, and its condition was similar too. Further digging revealed more of the same—no hominid bone, but fossils in the same state of preservation, all *in situ*. We could not be absolutely sure, but it was a pretty good bet: On the last day of our stay, we'd located the original resting place of our little *Homo* of Dik-dik Hill.

With that last-minute discovery, we could now understand in much better detail what had happened to that little hominid from the time of her death almost two million years ago to the moment Tim spied her broken elbow bone lying on the surface in the fading afternoon sunlight.

"This sand lens was probably deposited by a little rivulet draining into the Olduvai lake through the margin area," Tim told his students when we stopped for a midmorning break. "Judging by the assortment of anatomical parts we found so close together, I'd say the hominid died pretty close by, leaving her bones in one of these sand distributaries."

"From the weathering on the bones," I said, "they probably sat around on the surface for a year or two, maybe as much as five."

"Why aren't there any carnivore marks on the surfaces?" asked Berhane Asfaw.

"Who knows?" I said. "Maybe they missed it. Maybe there was better stuff around to eat than a scrawny three-foot hominid."

"Or maybe they took away the bones that they wanted," Tim suggested, "like that damn distal femur. There's plenty of grease for a hungry hyena inside a big articular joint like that."

"What if it *isn't* gone?" Gen Suwa offered. "We could dig deeper, follow the sand lens farther into the Gorge wall . . ."

"Sure, if we had another couple of summers and twenty guys to help us move dirt." Tim frowned. "If we'd turned up some of the hominid today, I'd say go for it. But the farther you move from the original find, the less likely your chances of finding anything."

"There's one thing I still don't understand," Gen Suwa said. "If it happened the way you've reconstructed it, why were the bones broken up so much?"

"To answer that you've got to look at what happened to the fossil after it was buried," said Bob Walter. "There are a lot of clays in this soil, which expand and contract, putting stress on the bones they surround. Calcite crystals and other minerals also grow in the cracks of the bone, wedging them apart. Meanwhile, you've got deposits piling up above. So while the fossilized bones might be shattered, the pieces are held tightly together by their surrounding matrix. A long time later, along comes some big erosional event, cutting down against the grain of all those layers. Later still the Olbalbal Basin sinks lower, cutting the Gorge deeper. Pieces of the Gorge wall

The possible fate of the Dik-dik Hill hominid after death. Top, left to right: The corpse of a mature Homo habilis *female dies near the Olduvai lake margin, vulnerable to scavengers. Over time, lake deposits and volcanic eruptions bury the skeleton deep under the surface. Natural underground forces shatter the bones, but the surrounding deposits hold them together. Bottom, left to right: Major erosional events cut down across the grain of the accumulated deposits. As the Gorge deepens, pieces of the wall break away. Centuries of wind and rain remove soft material unprotected by hard volcanic tufts, slowly bringing fossils and other denser materials to the surface again. Once exposed, the skeleton falls apart. Before all of it is washed away, the fossil-hunting team discovers the remaining bones on "Dik-dik Hill."* ILLUSTRATION BY DOUGLAS BECKNER

break away, like icebergs calving off a glacier. That's what Dik-dik Hill is, or was, before most of it eroded away.

"Where the wall is still intact," Bob continued, pointing up the slope, "hard caps like Tuff IC and ID protect the light material like the sand lens we found this morning. But where it is exposed, sand and other loose stuff washes out in the rains, or just blows away. The surface erosion slowly brings up the fossils and other hard stuff buried underneath. More of it rolls downslope. Which creates this hard 'desert pavement' you've been digging through all summer."

"Don't forget that the fossils are already shattered while still underground," said Tim. "So as soon as they hit the surface, they fall apart."

"So what you end up with is a couple of hundred tooth fragments, and only one complete tooth," said Gen, completing the answer to his own question. Tim grinned.

"Right. And when we get back to Berkeley, think of the fun you can have trying to put some more pieces back together."

Hans Schneider, the bush pilot, flew in just as we were finishing up lunch—too late to save me from trench duty, but still a welcome sight. Gerry and I got into the Land-Rover and drove out to the strip to meet the pilot. We found him tinkering with the engine, a rugged-looking man with smiling blue eyes and a disarmingly gentle manner. He was up to his elbows in grease, with Schubert's Unfinished Symphony booming from a tape deck in the cockpit. We introduced ourselves and told him what we were hoping to accomplish.

"You will not get good pictures now," he said. "The smoke from the grass fires on the plain is too heavy."

Hans would have to stay overnight. We all agreed that he and I would try to document the site on the way out the next morning, with a detour down to Laetoli before heading to Arusha. Tim, Bob, and most of the others would finish up work on the trench and then head to Arusha by Land-Rover bringing the hominid with them. Bill Kimbel and Alberto Angela would stay on at the Gorge and work Dik-dik Hill for a few more days.

We took off shortly after dawn, the nose of the plane pointed straight into the sun rising over Ngorongoro. It was a gorgeous morning, the air clear and cool. Hans motioned to me to put on the headphones. He slipped a cassette into his tape player, and the sunny opening theme of the Beethoven Pastoral Symphony tripped into my ears. Below me the Gorge looked like an emerald necklace thrown down on the spreading gold of the Serengeti, the light catching on the sisal spears lining the heights, carving shadows into the deep ravines. We banked over the Olbalbal Basin, dipped down into the main Gorge and breezed up its length, under the bronze cliffs of the Second Fault, past the stone hut sheltering the DK site, straight toward the Junction. In my ears the music swelled with optimism, the whole orchestra embracing that sunny theme in a bear hug, basses humming, flutes trilling, and for a moment it felt as if the airplane and the music were one, and I was being held aloft by the sheer, resolute joy of that melody.

As soon as we rounded the red-earth shoulder of the Castle I could see the Land-Rover parked beside the hominid site. The crew was already at work, puffs of dust rising from the sieve between Bill and Prosper. I took a couple of shots as we flew past. We were continuing up the belly of the Main Gorge when Hans abruptly pulled the plane up and banked to the north. He tapped me on the shoulder and pointed down. Ahead of us a small family of elephants was feeding below Granite Falls, a series of black cliffs near the western margin of the Gorge.

"The noise scares elephants," he said. "They panic, wasting energy that they badly need this time of year." He pointed down again, and I saw a pair of giraffes gliding across the plain toward the Gorge. "Them, they don't mind the plane," Hans explained.

We had seen plenty of giraffes around camp that summer, but the sight of the elephants surprised me. All summer long, preoccupied with the bones of the dead, we had been sharing the Gorge with these enormously obvious beings, without giving their existence a thought. How much other life breathed down there in the dry ravines? How many dik-dik pairs were domesticating the karongos with their scents and feces, how many lions and hyenas fed on the Olduvai night? As we banked for another run through the Junction, I saw a dozen oryx striding across the plain below Naibor Soit—graceful, big-bodied antelope with horns straight and long as lances. When the rains came they would be joined by a flood of other hoofed mammals washing in from the north—zebra, wildebeest, gazelle of all kinds, and the predators on their traces.

Two million years ago, when the Olduvai lake shone like a great silver coin in the middle of the plain, the mammal abundance would have been richer still. Below me I would see giant baboons, saber-toothed cats, pigs the size of buffalo, giant antlered giraffes, pachyderms like *Deinotherium* sharing the landscape with the animals still living there today. And amid all that heavy undulating mass of life there would be our little *Homo,* a naked two-legged thing with long arms and a body no bigger than a modern human child. How strange, I thought, that this little creature, this insignificance underfoot, had presaged an animal to come who would invent airplanes, write symphonies, and soar higher than a hawk above the ravines, a creature who would have the phenomenal presumption to wonder how he got so high in the first place.

For twenty years I had been trying to figure out how that transformation might have begun, but at that moment the mystery had me more baffled and enthralled than ever. Hans steered the plane back toward the Junction. Below, the excavations at FLK and FLKNN stood out like dead flesh, scars dug into the landscape by hungry, acquisitive hands. Around the bend, Dik-dik Hill bore a fresher wound. Tim was bent over an excavated square, worrying the earth with the pick end of his hammer. I kept my camera shutter clicking as we flew by. I thought about the effort we had expended there and the prize we had won for it: some three hundred chips of bone, which put together added up to one more morsel of information to feed this insatiable urge to know.

We took a few more runs over the site, and then followed the Side Gorge southwest toward Laetoli. My work at Olduvai was over for the year. Hans slipped another tape into the player, and Bruch's Violin Concerto sang through the headphones. Below me now lay the brush scrub of Laetoli. I shot my last two rolls of film and told Hans I was ready to head east toward Arusha. Instead he continued south along the Garusi river, flying only a few dozen yards above the dry riverbed.

"There is something I want to show you," he explained with a sly smile, and as he spoke the earth fell away beneath us. We had reached the Eyasi Escarpment, a sheer drop of three thousand feet. Far below, the parched bed of the salt lake shone with a pale brilliance, as if the crust of the earth as far as I could see was made of mother-of-pearl.

"Now we have a view, eh?" said Hans.

"Beautiful," I said.

Hans banked steeply and headed back to the east, skirting the south flank of Lemagrut, it shoulders pockmarked with grass fires. As we came into the lee of the volcano, a sudden downdraft hit the little plane with an audible thump. My head banged up against the ceiling and we began losing altitude. The sensation was less like falling than like being forcibly pushed down by an oppressive weight. Hans tried to pull the nose up but the plane kept falling, the engine screaming in protest and a half-charred ravine coming closer into focus. He gripped tight on the controls and gradually urged the plane sideways, away from the volcano. We had about five hundred feet to spare when the downdraft released us.

"That was not a scheduled part of the tour," Hans said when it was over. He was more upset than his words implied. We both had known people who had been caught in downdrafts like this one, and had not been so lucky.

As we rounded Lemagrut I looked back once more at the Gorge, now a mottled blemish on the edge of a plain stretching endlessly west. Through the years Olduvai had been such a fertile place for fossil hunters that sometimes it seems that early human evolution was an event staged at Olduvai Gorge. That is illusion, of course. Olduvai is a gift, a glorious accident of geology that punched a hole in the fabric of time and spilled a little of its contents. The same might be said of Koobi Fora, the Omo, the Hadar, the cave deposits of the Transvaal—all holes in the fabric. But time, and evolution, were happening everywhere.

The fossil we had found had revealed to us that the earliest humans were much smaller, much more primitive than we had imagined. But it could not reveal how such an animal managed to survive in the competitive world of the savanna. It could not explain why this diminutive ancestor had so suddenly become something far more human, in the transition to *Homo erectus*. It could not tell us what had become of its close cousins, the robust australopithecines. And it could not answer the question that I believed lay at the heart of the mystery: To what end had *Homo* evolved its appallingly oversized brain?

So where is one to look for answers? The best reply is "In better questions."

"The first step toward understanding is to stop assuming that humans win out in evolution *because they are human*," says Robert Foley, a young paleoanthropologist at Cambridge University in England. An archeologist and ecologist by training, Foley believes that far too much homage has been paid to the uniqueness of humankind, at the expense of understanding our genesis as a product of evolutionary forces shaping a whole community of other organisms at the same time. Our self-fascination has led us time and again into the same tautological trap: We look for some trace of uniquely human behavior in the ancient record, and once it is found, we point to it proudly and declare that we now understand the origins of humanity. If you define man as "the technological animal," for example, then of course you have only to find the first evidence of tools, and *presto*, you have discovered the beginning of man.

As Lew Binford would say, that is *post hoc* thinking, studying the past with one eye in the mirror. Toolmaking. Bipedalism. Brain growth. Hunting. Food sharing. All of these things *did* develop at some time in the human career, but we'll never comprehend how or why or what for unless we stop thinking backward from the *result*. Only when we strip early humans of their future can we begin to understand the secrets of our past.

I want to talk about hominids. But I first have to review a few points about *species*, whether they be hominids, hippos, or hibiscus plants. Point number one is easy: All species are adapted to the habitats they live in. You cannot ask a hippo to rise from her watery niche and populate the grassland, any more than you can expect a desert flower to thrive, transplanted into the jungle. Some species may be specialized to live in *one* kind of habitat (marsh lilies, penguins), while others are adapted to a range of habitats (cockroaches, us). But you still won't find any species dwelling for long where it is not physically or behaviorally adapted to make a living. It needn't be perfectly in tune with the environment; it need only be adapted *enough* to reproduce another viable generation of hominids, hippos, or hibiscus.

The second point follows from the first: Species tend to be very *conservative*. In most circumstances they do not show much change generation after generation, and for good reason. So long as their habitat remains fairly stable, those members of the species who are best fitted to the prevailing conditions will be those most likely to survive and reproduce, passing on the same old conservative traits that allowed them to succeed. If and when new species arise in stable habitats, the change is likely to be so subtle, so gradually paced, that the dividing line between the ancestor and the descendant is quite impossible to detect. Some theorists, including many who subscribe to the well-known "punctuated equilibrium" model of evolution, believe that new species cannot originate *unless* there is a substantial change in the environment first. I do not think this has to be true—but I do agree that it is in times of environmental instability that evolution really begins to heat up.

Point number three: While the environment outside provides the impetus for change in a species, the response that the species makes will be constrained from within. The biology of every toad, daffodil, and tsetse fly is a legacy of its evolutionary history. That heritage is inescapable; whatever anatomical or behavioral innovations are offered by the next generation of toads, they will always be Variations on the Theme of Toad. For this reason alone, you cannot focus a narrow beam on the earliest *Homo* and understand what it was doing to become "human," unless you consider too its australopithecine and primate pasts.

My fourth and last point: The creation of a new species requires that a population of its ancestors be blocked from reproducing with others outside the population. Usually this reproductive isolation comes about geographically, resulting in what is called allopatric speciation. Allopatric means "in another place." Brought to prominence by the Harvard evolutionist Ernst Mayr, the allopatric theory holds that new species usually originate in small populations isolated on the periphery of the species's ancestral range. Closer to the center of the range, the large, stable populations enjoy a sort of built-in genetic inertia; the conservative traits of the many tend to obliterate the evolutionary innovations offered each generation by the few. In the small, peripheral populations, on the other hand, random genetic variations are more likely to catch hold and express themselves in the next generation. Natural selection works with a sharper edge in these marginal habitats, since they offer unfamiliar environmental challenges.

Now let's work our way back to hominids. As Hans Schneider and I flew east over Ngorongoro, I peered down through the mist and could just make out the lush greenery covering the volcano's slopes. Ngorongoro is an island of forest in the surrounding plain. During the early Miocene, some sixteen to twenty million years ago, this kind of forest would have stretched uninterrupted across the Serengeti and all the way to the west coast of Africa, and northward across the Sahara Desert, blanketing virtually all of Africa and much of Europe and Asia as well. Living under that endless canopy were dozens of species of hominoids, among them the ancestor to all existing apes. The environment was stable, and for an arboreal, fruit-eating ape, making a living was relatively easy.

But nothing is stable forever, least of all the planet itself. Unbeknownst to the apes, the continents were sliding around under their feet. Long before the Miocene, Africa, India, South America, Australia, and Antarctica had all been part of one megacontinent that scientists call Gondwanaland. Over hundreds of millions of years the movement of the tectonic plates making up the earth's outer shell had been pulling the gigantic landmass apart. By sixty million years ago, only Australia and Antarctica were still connected, and Antarctica enjoyed a relatively temperate climate. But when the two other continents broke free and drifted north, an ocean current developed around Antarctica that effectively blocked warmer water from reaching its latitudes. (Today this circum-Antarctic current is the strongest on earth, transporting some twenty million cubic meters of jater per *second.*) The isolation of Antarctica led to massive buildup of the continent's polar caps, causing a significant drop in the level of the world's oceans.

Until just a few years ago, it was thought that the buildup of ice at the earth's poles led to increased rainfall and vegetation in the tropics. We now know that exactly the reverse is true: Global cooling leads to more arid conditions around the equator. And so, gradually, the Miocene rain forest began to break up. From the nature of pollen preserved in soil, and from the kinds of animals we find inhabiting the regions, we know that open woodlands and even some grassy expanses were gaining territory at the expense of tropicalforest.

Then came a deeper catastrophe. With the help of some local tectonic movements, the drop in sea level began to squeeze shut the narrow channel between the Mediterranean Sea and the Atlantic Ocean. As the bottleneck between the two seas grew shallower, enormous quantities of salt began to accumulate in the Mediterranean. Between six and a half and five million years ago, the Mediterranean trapped in its shrinking volume as much as 6 percent of the world oceans' salt.

The "Mediterranean salinity crisis," this quirky aberration in the simple chemistry of the ocean, may have been a kind of climatic trigger to the human career. Fresh water freezes at a higher temperature than salt water, and a world ocean 6 percent fresher than normal is no exception. The result was a sudden, accelerated buildup of sea ice around Antarctica, further lowering the sea level to the point where the Mediterranean basin was left high, dry, and a mile deep in salt. The earth's temperature took an abrupt plunge, a much steeper drop than the gradual cooling trend that had prevailed for millions of years.

Think back to the theory of allopatric speciation—new species erupting on the periphery of their ancestral habitat. Now think about a forest breaking apart, becoming not one far-flung mass but a scatter of smaller patches, ever more scattered, ever smaller. In some areas, the forest might be reduced to the greenery clinging to the mountains. Ngorongoro beneath us. Mount Meru north of Arusha. Kilimanjaro. Mount Kenya. And as the forest goes, so go the species who depend upon it. Instead of one large ancestral range bordered by peripheral areas, the hominoid apes, along with everybody else, ended up in many "islands" of the old preferred habitat, *each* isolated, *each* surrounded by the kind of iffy, marginal zone where rapid evolution is likely to occur.

Elisabeth Vrba, a South African paleontologist now at Yale University, has an elegant way of describing the choices* facing a

* Biologists use words like "choice," "strategy," and "solution" to describe the responses of species to their environments. But species, including our own, do not make conscious decisions about the course of their evolution. The language of consciousness is used merely as an analogy for the way natural selection and adaptation work.

species whose habitat is breaking apart under its feet. Elisabeth says the species can take any one of the three paths offered by the "Hindu Triad" of gods. It can follow the way of Vishnu the preserver and migrate to a new area, where the old environmental conditions still prevail. With this option the species need not give rise to new forms, nor even undergo much change within itself. The second choice is to stay put geographically, and be led by Brahma the creator, bringing forth new species better fit to the conditions that have taken over the home region. The third choice is to fall into the waiting arms of Siva the destroyer, and become extinct.

From what we can see in the fossil record, by six million years ago most of the dozens of hominoid species which once inhabited the rain forest had followed Siva the destroyer out of existence. Of those few that remained, most followed Vishnu to the patches of forest resembling the original habitat. In the late Miocene, coinciding with the Mediterranean salinity crisis, one ape took the path of Brahma, and a species it spawned walked there on two legs. It became adapted to more open habitats, where a bipedal primate could forage with success at least equal to that of a quadrupedal one. Not everyone thinks that the first hominid developed bipedalism *because* it was a more efficient way of coping with open landscapes. Owen Lovejoy, for example, is convinced that Lucy's ancestor was bipedal while still adapted to the Miocene forests. But whatever the prime motivation for bipedalism, the fragmentation of the forest gave this novel trait room to establish itself: an evolutionary innovation of a small group of prehominids, tucked away out of reach of their four-legged ancestral stock.

There is no fossil I would rather find than one plucked from that population of transitional bipeds. But the chances of that wish coming true are vanishingly remote. The transition probably would have taken place too fast and in too restricted an area to be viewed from one of the geological "windows" represented by the fossil sites of Africa. My guess is that the end result of that innovation was *afarensis*—or something very close to it. The oldest signs of *afarensis* in the fossil record are the fragments of frontal bone and femur that Tim and Desmond Clark found on the expedition to the Middle Awash in Ethiopia in 1981. These have been dated to be roughly 4 million years old. Meanwhile the *youngest afarensis* from Hadar carries a date of approximately 2.9 million years. That means that the species survived at least a million years, perhaps far longer. By any-

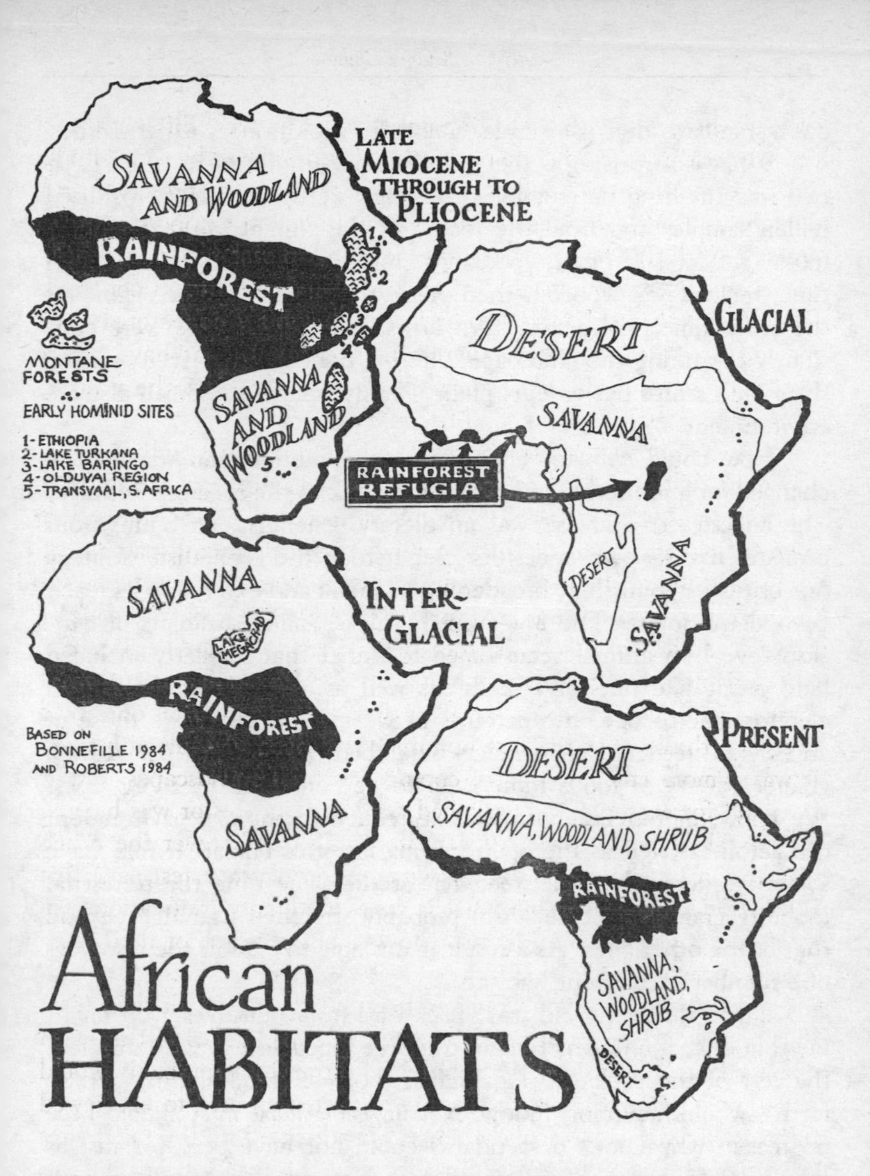

African HABITATS

The pattern of change in the environments of Africa helped to shape the evolution of the hominid lineage. In the late Miocene, the tropical rain forest that once dominated the continent began to give way to savanna and open woodlands. Over the next 10 million years, polar glaciers exerted a profound effect. Cooler temperatures and lower rainfall during glacial periods shrunk the forest into scattered patches, isolating the species within them. The rain forest expanded again in the warmer, wetter periods between glacial advances, bringing together species that evolved in isolation.

one's standards that is a stable, successful evolutionary experiment.

What is surprising is that the Pliocene environment of East Africa was anything but stable. From what we can tell from fossilized pollen samples and from the record of the animal inhabitants, the mosaic of forests, open woodlands, and savanna flickered through time, regions first wooded, then open, then forested once again. At the same time, volcanoes, lava flows, and earthquakes were constantly reworking the landscape, altering the course of rivers, raising sheer fault scarps out of level plain. Hardly what I would call a stable environment.

How could *afarensis* endure such commotion and remain unchanged for a million years? For a clue, look to the species's heritage. The ancestor of *afarensis* was an dietary generalist, an omnivorous primate, like its own ancestors. But it took the generalists' strategy one critical step further, broadening its menu and extending its reach beyond the forest. The low-cusped, thick-enameled molars of *afarensis* reveal an animal accustomed to a diet that regularly included hard items like nuts and seeds, as well as softer fruits. Under an electron microscope, the patterns of wear on its teeth are those of an eclectic feeder, their surfaces scratched, pitted, and flaked by a greater variety of foods than seem to be consumed by living apes. We know, moreover, that *afarensis* frequented the dry environment of Laetoli, as well as the open canopy forest of Hadar. It may have sometimes foraged in the trees, but at the same time the terrestrial mobility granted by bipedalism probably stretched its range beyond that of the other apes. As a result it was able to exploit the resources of a number of different habitats.

If you think the old way, backward from ourselves, you might say that once a uniquely human trait like bipedalism was established, the rest of the package—big brains, tool use, food sharing and so forth—would inevitably follow. *Nothing is inevitable in evolution.* I see no reason why Lucy's descendants could not have persisted just as they were for another million years or more, well-endowed generalists in a fairly rich habitat, apes with brains hardly bigger than other apes, walking on two legs because it suited their *present* circumstances, not because it augured some glorious future. But *afarensis* did change. Between three and two million years ago, East Africa became an even harder place to make an easy living, perhaps quite suddenly. And *afarensis,* like its own ancestor, was forced to choose between the paths of Vishnu, Siva, or Brahma.

I've talked about the changes wrought upon Africa by the global temperature plunge in the late Miocene. Recently, Elisabeth Vrba drew my attention to a rash of new evidence pointing toward another climatic catastrophe later on. Deep-sea sediment cores record a surge in polar ice between 2.5 and 2.4 million years, perhaps marking the first glacial advance in the Northern Hemisphere. Pollen studies on land deposits tell the same story: In Holland, palm forests give way to open steppes; in Colombia mountain forests wither into plain. Pollen samples from the Omo in Ethiopia show an abrupt shift from woody plants to grasses and low shrubs. The Omo's record of microfauna—rodents and the like—neatly parallels the vegetative switch, with forest-living forms giving way to arid-adapted types between 2.5 and 2.4 million years ago. Most recently, deep-sea cores from the ocean off West Africa are coming up full of dust at the same point in time. From modern studies, we know that dust settles in the ocean where there is desert on the adjacent land.

Elisabeth's own work on African bovids comes up with the same conclusion, and in spades. Bovids are especially good indicators of evolutionary change in Africa simply because there are so many of them. Two and a half million years ago, the bovid family underwent an explosion of new species well adapted to savanna conditions: the hartebeest and wildebeest, the gazelles, impala, springbok, and similar forms.

"All the continents are saying the same thing," Elisabeth told me, with obvious delight. "The climate is changing in some places, with dramatic force. The cause of the change is not clear—perhaps tectonic movements closed the Isthmus of Panama, reshuffling the circulation of currents in the Pacific and Atlantic oceans, which in turn affected temperature and precipitation patterns. But that it *did* happen is almost beyond doubt."

According to her "turnover-pulse" theory, this sudden shift in the global climate might have sent a surge of extinction and speciation through the food chain. The pulse would hit the early hominids too, possibly triggering the appearance of *Homo*. We do not have any direct evidence for *Homo* that early in the fossil record—not yet—but we do have the next best thing. The discovery of those very primitive stone tools near Hadar matches the date exactly. Equally important is the *absence* of tools any earlier. Stone tools are not time-fragile like fossils. If we look for them, and we can't find them, then they probably weren't there to begin with. If Elisabeth Vrba is right, and I think she is, the *Homo* emerged in East Africa some 2.5 million years ago, one among many species struggling to adapt to radically altered conditions.

"What is true on a personal level is true for evolution too," she says. "You are not going to change much if you sit around doing the same old thing in the same familiar habitat. But when you are thrown into situations of stress, of unfamiliar risk, you are pushed to change. We are the product of catastrophe."

We have to look a little more closely now at this new world facing "Lucy's children"—the descendants of *afarensis*. So far, I have been talking about habitats in rather two-dimensional terms, sliding in words like "forest," "woodland," and "savanna" as if they were painted backdrops to the real action taking place stage front. That is convenient shorthand, but it reinforces the notion that human evolution was somehow special and distinct. Humans did not evolve *in* habitats; they were *of* them, like other animals, the way that molecules in an electrical wire are part of the energy running through. An animal's habitat is the temperature of the air next to its skin, the rainfall dampening the dust beneath its feet and its cycle through the seasons, the height of the mountains and depth of the winters, the salt in the earth and the oxygen in the air. By day a habitat sucks up energy from the sun, by night it gets hungry and eats its own tail. A habitat is *alive*. It moves. It runs away from you or snaps at your hooves. It gets to the food before you do, or it paces nearby while you eat. It may steal your meal, your portion of the energy, from under your nose, or it waits in your gut and steals it later. A habitat is dynamic. It is more than the physical environment and the sum of the organisms that live and die in its boundaries; it is the conflicts and interdependencies that bind them together. A habitat doesn't stand still while you evolve to master it. It evolves too.

"Savanna" is a broad term embracing everything from gallery forests lining rivers and lake margins to semi-deserts with scarcely a tree in sight. In between are moist woodlands, drier woodlands, bush country like Laetoli, and vast grasslands punctuated by the occasional solitary acacia. In spite of this diversity, there are conditions that all these habitats share in common. Savannas do not support much "primary production"—the foundation of plant material that carries the hierarchy of animal life in the food chain above. In spite of the apparent barrenness of the bottom line, however, savannas often show a great abundance of animal life. Anyone who has had the good fortune to witness the annual, horizon-spanning migration of wildebeest in the Serengeti Plain knows what the word "abundance" can mean in this context. The key to this mammalian munificence is *grass*. Grass is low-quality food; it does not hold much protein, and what is there is locked up in tough cell walls. But on the plus side, grass regrows quickly, so supplies are constantly renewed. Hoofed ungulates like the wildebeest, zebra, and gazelle "solve" the problems posed by grass feeding through behavioral routes (eating all the time) and anatomical ones (specialized digestive equipment capable of breaking down the cellulose in grass). These prey species naturally attract predators. Savannas are also home to a few primates and pigs that feed on high-quality, scarcer plant foods.

Our ancestors did not enter the plains like Oklahoma land-rushers racing to stake a claim in virgin territory; they were thrown into an environment already populated with highly successful savanna veterans. Like any species trying to carve a niche for itself in a new habitat, the hominids would have to sniff around the edges of the energy system, looking for food resources left unexploited by more established species. I doubt very much that *Homo* joined the ranks of the carnivores and took to hunting. Tools or no tools, how is a fruit 'n' nuts kind of generalist supposed to muscle a niche for itself between a five-hundred-pound lion and a pack of sly hyenas? The first hominids on the savanna did not have to compete with carnivores—they had only to stay out of their reach. Their real competitors would be other high-quality plant-food eaters, like the baboons, monkeys, and pigs. Like these other species, the hominids would have to come to terms with a fundamental fact of savanna habitat: Premium resources are rare and scattered. To keep alive, you will have to forage farther, with greater reward attending whatever extra effort you put in.

But that's only half the problem. The savanna confounds the search for food in the dimension of *time* as well as space. People tend to think of savannas as dry, which is only partially true. Some savannas receive as much as 100 inches of rain a year. What matters, however, is the *timing* of this rainfall. The most distinguishing characteristic of the savanna environment is seasonality. The rain comes in bursts. The periods in between—the dry seasons—may last anywhere from two and a half to ten months. The dry-season savanna is not just a thirsty version of the wet-season savanna. It is a different place altogether. In the dry season the grass shrivels to a useless brown mat. Where there was food, there is dust. In low elevations the heat is intense, and even on the relatively cool plateaus like the Serengeti, there is little shade to escape to from the sun. The herbivores, dispersed during the wet season, either migrate away or flock to permanent sources of water. There many die of starvation, or fall victim to the predators who await their coming. Animals adapted to higher-quality plant foods suffer just as much. They stand in double jeopardy: As each parcel of their range grows increasingly impoverished, the drying up of water holes keeps them bound to a smaller and smaller foraging area.

To every animal on the savanna, the dry season represents a chasm of lean times that has to bridged year after year. The first thing an animal does when faced with a lapse in resources is cut down on activity to save energy. Not many mammals have evolved the radical extreme of hibernation, but most show a falling off of play, socializing, and reproducing during the hard months. But rarely is energy saving alone enough to allow survival until the rains return. Each animal has to make a crucial decision: either to start eating lower-quality foods like mature leaves, seeds, and stems that are still abundant in the dry season, or to find ways to win access to higher-quality foods previously out of reach.

I think it was this dry-season "decision" that really forced the branching in the hominid line. In response to the seasonal stress, some populations of *afarensis* developed the customized chewing equipment—those massive jaws and heliport molars—that would enable the robust australopithecines to cope with the tough, low-quality vegetation available in the dry season. But another population took a second route. Building on their phylogenetic history as resourceful, omnivorous primates, the ancestors of *Homo* began to poke around in the dry season looking for new opportunities to get at out-of-reach foods with high payoffs. They did not use specialized jaws and teeth to get at these prized goods. They used their heads.

Brains are wonderful things. Speaking from the admittedly biased point of view of a large-brained primate, I would say there is no better solution to one's environment—no claw so sharp, no wing so light that it can begin to bestow the same adaptive benefits as a heavy ball of gray matter. What a big brain gives you is *flexibility*. With a brain you are no longer a creature dependent on the moment, doomed to programmed responses that may or may not be equal to the task posed by the new challenges you face. With a brain you can learn from past situations, foresee the advantages of two different courses of action, and calculate the best alternative. Big, powerful, calculating brains are so useful that at first glance you might wonder why so few species have developed them—the primates, some cetaceans, to some extent elephants, a few carnivores.

The answer is that their cost is prohibitive. Modern humans spend about 20 percent of their metabolic energy keeping their brains running, as opposed to about 10 to 13 percent in a relatively cerebral nonhuman like a monkey. In human infants and children under four, the brain devours closer to 50 *percent* of total body metabolism. Half the blood supply pumping for one organ. The brain's demands are relentless. They go on even when you sleep, even when you watch reruns of television shows you thought insipid the first time around. Brains are also freeloaders. They hold no energetic reserves themselves. If the supply of oxygen and glucose is cut off, the brain quickly collapses into stupor, with vast and irreversible damage to its tissues within minutes. When you add up all its costs, a brain makes evolutionary sense only if you really need what it provides.

What early *Homo*'s brain provided was *access*, the means to break through the inhibiting protections around food pockets and scoop out their rich, sequestered rewards. Leaning on sheer versatility, they eked out a living where none was there to be found, borrowing a little from the fringes of one species's niche, taking the crumbs that fell from the mouth of another. Some of these rich nutritional sources would be virtually invisible. The African savanna is full of "drought-evading" plants that keep a large part of their energy reserves stored underground. In the dry season, when edible vegetation aboveground grows scarce, the tubers, roots, bulbs, and other nodulous larders of the drought evaders make up a huge proportion of the total plant biomass—all of it available to rooting pigs or clawing bears, or to a primate clever enough to conceive the idea of a digging stick. Other dry-season energy sources are highly visible, but more difficult to catch.

"One of the best solutions to getting through hard times is to eat somebody else who has already taken the trouble to come up with a different solution," says Lew Binford.

When Lew told me that, he was referring to the hunting behavior of Ice Age humans a mere forty thousand years ago. As you know from the last chapter, he scoffs at the notion of *habilis* as a hunter, and by and large I have to agree with him. But to rule out that early *Homo* was eating *any* live prey is too extreme. I find it hard to believe that a large-brained, undoubtedly social animal like *habilis* would eschew a rich food source that some modern nonhuman primates regularly exploit. When Jane Goodall first observed chimpanzees hunting twenty years ago, it was assumed to be a rare, isolated phenomenon. Through her subsequent studies and those of others, we now know that the hunting of other animals, often planned, deliberate, and highly cooperative, is characteristic of many chimpanzee communities, though some do it more than others.

And chimps are not the only primate killers. Humans aside, the most predacious primates known are the members of the "Pumphouse Gang," the group of olive baboons in Kenya that have been under the scrutiny of Shirley Strum of the University of California at San Diego for over fifteen years. Strum's baboons have developed sophisticated strategies like relay chases and long-distance pursuits to capture young Thompson's gazelles and other small ungulates. In one year they averaged one kill *every twelve hours*—hardly what you would call gentle herbivores. Granted, the Pumphouse Gang might be an unusual extreme. But they stand at one end of a very suggestive trend: In both chimps and baboons, the groups that inhabit savanna environments include a much greater percentage of meat in their diets than do their forest-dwelling counterparts, and they hunt most during the dry season.

Oddly, today few chimps or baboons appear to avail themselves of that other method of obtaining meat—scavenging. Early *Homo* was not so choosy, and according to Rob Blumenschine, now at Rutgers University and a member of the Institute's research team in Tanzania, the opportunities were there to be exploited. In 1983, Rob conducted a study of what happens to a carcass after death on the savanna—how much the killer eats, what kinds of scavengers come to sup upon the carcass later, and what they leave behind for others—to see if there would have been anything left on a regular basis for a hypothetical early hominid to exploit.

After several rather messy seasons of fieldwork, often up to his elbows in half-eaten wildebeest carcasses, he concluded that a "scavenging niche" might indeed have been vacant in the Plio-Pleistocene, ready for hominid occupancy—but only in certain habitats. Among all the carnivores on the savanna, only the hyena eats all of a carcass, leaving nothing of worth behind for subsequent scavengers. Rob reasoned that the places where hominids might have profited from the kills of predators would be where hyenas weren't around to consume everything first. Hyenas prefer open plains or lightly wooded habitats. In the infrequent times when Rob encountered them in more densely wooded areas, such as the forest fringes of lakes and rivers, they seemed "skittish," apparently wary of lions. In these areas the marrow and head contents of a lion's kill often sat untouched until they finally began to rot. Rob's study did not prove that hominids were scavenging—only that the opportunity may have existed. From sites like Olduvai Gorge and East Turkana, however, we know that the lake and river margins seemed to be a preferred habitat of *habilis,* and that they appeared to be using stone tools to break open bones.

The lasting testimony of *Homo*'s heightened foraging access is the ancient tool, used to coax tubers out of the dead ground, deflesh carcasses, and hammer marrow out of barren bone. But because an activity requiring intelligence just *happens* to have archeological permanence does not mean that toolmaking is the sine qua non of the human mind. The ecological opportunity I see opening to *Homo* extends far beyond what we see expressed in a chopper or flake. *Knowledge,* not its manifestation in stone, is what pried out the hidden scraps of energy dotting the ancient landscape. The first humans survived the dry season by knowing their habitat in intimate detail. They were opportunists armed with information, and probably relied a great deal on each other to communicate knowledge of food patches separated by long distances. They knew from the shape of a dessicated stalk what pithy tumescence lay in the ground beneath. They knew the promise implicit in the circling of a vulture, a monkey's scream, the urgent trilling of a honey guide bird as it led them to a bee's nest. They knew that adult antelopes, while impossible to catch, sometimes "park" their young in grass and go off to browse. Though invisible to the motion-sensitive eyes of a carnivore, a generalist with a primate's color vision could pluck the infant like ready fruit—if he knew where to look.

Obviously, the big-brain flexibility evolved to cope with dry-season stress would be available in the wet season too. The early humans understood that the pattern of food on the landscape shifts in complex but predictable patterns: that fig tees in the gallery forest ripen at specific times, that toxic berries can be separated from the edible ones that mimic them by shape of leaf or feel of bark, that the termite's nuptial flight attends the first drops of rain and for a few days fills the air with food. Bird's eggs are rich morsels too *if* you can get access to them, perhaps by knocking down a nest with a well-aimed stone, or remembering what kind of cover a nesting pair used in earlier years. Later on in the wet season, when the fledgings take their first independent flights, there may be some who fail the test, their panicked thrashings revealing that they are ready to be fished out of the avian gene pool. What set *Homo habilis* apart and defined the human niche was not hunting, not scavenging, not digging sticks or any tool per se, but the apprehension of possibility in an unpromising landscape. An animal on the make.

On the flight east from Olduvai I looked down upon the Kainam Plateau, wearing its pallid late-summer complexion. We soon reached the escarpment above Lake Manyara, and then, abruptly, Manyara itself, a dark ribbon of forest threaded between the lake and the highlands to the west. The fringes of the lake were dappled with bird flocks and slow mammalian motions; I could see elephants browsing in fever trees near the shore, and in the shallows, an archipelago of wallowing hippos. Nearby a herd of zebra bulged around a stream. I could not see the lions sleeping in the trees, the warthogs quickstepping in the thickets with their purposeful gait, or the other small creatures picking out a living in the sunlight or curled in their burrows, waiting for dark. But they were there too, and if you took away the hotel perched on the escarpment, the cultivated fields, the shadow of the plane and other signs of later men, it wasn't so hard to imagine *habilis* in this world after all. There would be no stone dwellings by the lake, no little families waiting in the vestibule for Daddy Homo to sharpen his chopper and carve a thick steak off his latest kill. But there would be small, brown figures darting among the shadows of the umbrella thorns, drawn to the lakeside because that was where the energy resides—in the trees, in the ground, in places where no one else thinks to look. They could scrounge out a living that way, if they relied on each other and assiduously avoided known dangers. Early *Homo* survived, barely, without a sheen of special destiny protecting its progress toward dominion. Personally, I find that more impressive than any of the old myths.

And what of the hominid siblings, the robust australopithecines? They would be down there too, of course. Because the path they took led eventually to extinction, the robusts are often thought of as innately inferior to *Homo* from start to finish. Nonsense. Judging from the sheer number of fossils we have found, *A. robustus* and

A. boisei were very successful species in the early Pleistocene, often outnumbering their *Homo* counterparts by two to one. By definition, the adaptive strategies of the two lineages would have overlapped in the beginning, including the ability to access food resources through ingenuity. Chimpanzees use tools effectively, and it is almost inconceivable that the australopithecines would not have done so too. From a level at Swartkrans associated with *robustus*, Bob Brain has recently unearthed long-bone splinters that appear to be polished and may have been used as digging sticks. *Homo* is represented by only a few fossils at Swartkraus. Is it possible that *robustus* was using the digging sticks? Perhaps, but we cannot be certain.

Another prevalent myth has it that the robusts were driven to extinction by competition from their smarter *Homo* relatives. That idea is unlikely, at best. The robust and *Homo* lineages would have competed for resources intensely near their point of divergence, when their ecological niches butted up against each other. But as time progressed, their differing strategies for coping with seasonal stress would have taken them ecologically farther apart. The robusts did not disappear until just after one million years ago, long after *Homo habilis* had evolved into *Homo erectus*. Robert Foley speculates that if the robusts *were* squeezed out of existence by foraging competitors, the pressure would have come not from *Homo erectus* but from the bovids and other herbivores who could better exploit the low-quality plant foods to which the robusts had become adapted. Alan Walker sees a different role for *Homo* in the extinction of the robusts: as a member of an efficient new guild of predators. Meanwhile, Elisabeth Vrba points to the possibility of another episode of global cooling near nine hundred thousand years ago. She suggests that the demise of the robusts might be understood in the context of another massive wave of mammalian extinctions and speciations. The truth is we do not really know why the robusts faded from existence. But all species do, eventually.

There are still a great many questions to be answered about Lucy's other child as well—the *Homo* lineage. For starters, the controversy surrounding *habilis* remains unresolved. *Homo habilis* was an ill-defined taxon when Leakey, Tobias, and Napier named it back in 1964, and none of the dozen or so fossils found since—not even the famous 1470 skull from Koobi Fora—have made things any clearer. Some anthropologists, such as Chris Stringer at the British Museum and Bernard Wood of the University of Liverpool, have recently ar-

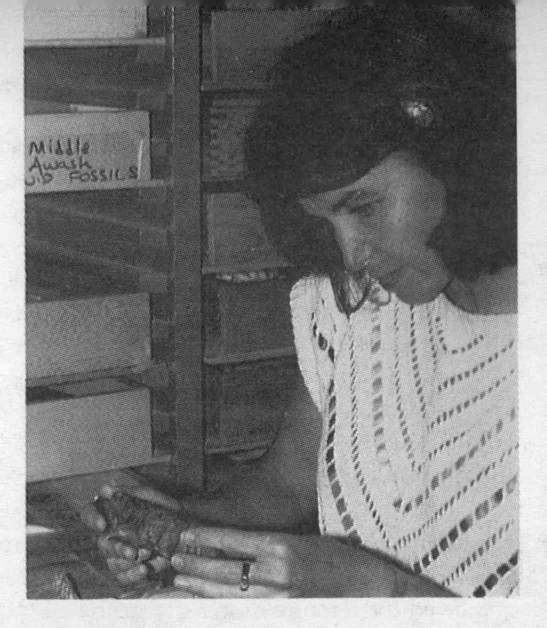

Elisabeth Vrba examining afarensis *fossil vertebrae.*
DONALD JOHANSON

gued that there is too much variation among the accumulated *habilis* specimens to fit comfortably into a single species, and that there may in fact be two species represented in the fossil sample. The diversity they point to is most evident in cranial capacity, which ranges from just over 500 cc to about 750 cc—too great a reach, perhaps, to be spanned by the single name *habilis.*

Stringer's and Wood's interpretations imply more than just a little taxonomic housecleaning. If there *were* two species of *Homo* living contemporaneously in East Africa, then probably only one of them could be an ancestor to *Homo sapiens.* Taking that thought one step further, you end up with a sort of shadow human lineage, evincing many of the characteristics we thought to be ours alone, then disappearing from the fossil record. It is a fascinating possibility, and unlike the arguments for two species roped under the name *afarensis,* deserves to be considered very seriously.

At this time, however, I do not think we have sufficient evidence to justify splitting the *Homo habilis* sample in two and naming a new *Homo* taxon. J. Miller of the University of Southern California recently showed, for instance, that large samplings of gorillas, orangutans, chimps, and humans show a range in cranial capacity equal to or greater than that of *Homo habilis*—and nobody doubts that these are single species. My working hypothesis is that the variations found among the *habilis* specimens have a great deal to do with sex, geography, and time. The big individuals are probably males, the smaller ones mostly females. The sites where the specimens have been found are as far apart as five hundred miles and may cover as much as three or four hundred thousand years.

Which brings us back to the mystery so potently underscored by the Dik-dik Hill hominid. If this little skeleton is a typical *habilis* female, then the transition to *Homo erectus* in that critical time-slice between 1.8 and 1.6 million years ago was sudden and dramatic. We do not know why that transformation occurred, but we do have a few leads. So far there is no indication of another global climatic trigger at that time, but Richard Hay's geological work tells us that the local Olduvai habitat, at least, dried out about 1.7 million years ago. The lake shores shrank in. The fauna switched from species comfortable in moist conditions to those adapted to open grassland. The same change can be read farther north in the Omo's Shungura Formation, and perhaps in South African cave deposits too. A big-bodied creature like *erectus* could exploit the larger home range necessary under such conditions, and would be in a better position to face predators in the open. A big body would also be needed to metabolically support a bigger brain, since neural activity consumes considerable energy. Certainly a lineage already committed to intelligence as a foraging strategy would use that flexibility as it continued to adapt to new habitats farther down the evolutionary road.

That swollen cerebrum would bring with it some inflated costs. In addition to its even greater need for water, the absolute energy requirements of a big *erectus* individual, like the five-and-a-half-foot skeleton of an adolescent that Kamoya Kineu found in 1984, would have been much greater than those of the tiny Dik-dik Hill hominid. Alan Walker believes that beginning with *erectus*, our ancestors became much more dependent on live prey, so much so that the species might be described as "a stalking species of hunter, as men are today." The hindsight perspective implied in that statement bothers me, but I suspect that *erectus was* relying on more meat, especially in the dry season. I do not see how a dietary generalist with that kind of body bulk could have survived without leaning more heavily on animal protein than its compactly proportioned predecessor.

And of course the Acheulean tools appear in the archeological record along with *erectus*. The temptation remains to put man and tool together, and see the advance of technology as the driving force in the evolution of *erectus*. But be careful—we might start reading the text backward from the conclusion again.

"I'm sick and tired of hearing people say that '*erectus* equals

Acheulean', as if there was a gene at position twenty-two on the fifteenth chromosome programming that particular hominid to start making hand axes," says Tim White. There is no reason to assume, in other words, that the biological transition to *erectus* was somehow *dependent* on an advance in technology.

"Tools were adjuncts, not the means of adaptation like they are today," adds Lew Binford.

To my mind, it was probably the pressure of an increasingly impoverished habitat working upon the hominid heritage that brought forth big *erectus*, much as it had earlier initiated our divergence from the robust australopithecines. But that cannot be the whole story, not by a long shot. Environmental stress may have provided the "kick," but other influences were in play as well, amplifying the direction, quickening the pace, reinforcing each other. It makes no more sense to look upon *Homo* as purely an ecological entity than it does to consider the invention of stone tools in isolation, or to attribute our evolution to some one, uniquely human behavior. Simple answers might be satisfying, but they will not lead to the truth.

Before descending from this flight over the landscape of our beginnings, I have to bear witness to one evolutionary "amplifier" in particular—the role that hominid society itself must have played in shaping our intellects. Generally speaking, paleoanthropologists today tend to shy away from speculations on early human social behavior. We relish hard evidence, and start to squirm in its absence. (Not once during a field expedition have I glanced down to see a fossilized social interaction lying at my feet.) But to ignore the pressure of society upon our ancestors just because it leaves no permanent record would be ridiculously shortsighted. While my colleagues and I have been busy scanning the ground for physical testimony to the human past, primatologists and animal psychologists have been looking for other clues in the social lives of our closest living relatives. What they have been finding there most recently bears directly on the origins of our intelligence, and is too disturbing to be ignored.

As Hans Schneider brought the Cessna around to land on the Manyara strip, I looked west one more time toward Olduvai. Beyond the Serengeti, over the horizon and across Lake Victoria lay the Virunga mountains, where Dian Fossey had so long studied and lived with the mountain gorillas. In the early 1970s, a young British psychologist named Nicholas Humphrey spent some time working in the

Virungas with Fossey. Like other researchers before him, Humphrey was struck by the apparent ease of gorilla life: They had no natural predators other than man, and to eat, they had only to reach out and grab a handful of leaves. Humphrey knew that gorillas, like other apes, show amazing intelligence when put to the test in laboratory experiments. Therein lay a powerful paradox, for if intelligence evolved in response to environmental pressure, why was so unchallenged an animal so manifestly smart?

In an essay called *The Social Function of Intellect,* Humphrey answered that society itself was the driving force behind primate intelligence, and even more so the spur to the astonishing intelligence of man. Society acts as a sort of double-edged knife to shape the intelligent being. On one side, a social group provides the context for learning about the nature of the habitat that every young animal will need for its future survival. Chimpanzees, for instance, do not come into the world knowing how to fashion a twig into a probe that can be used to fish termites from their nest; nor are they likely to discover the trick on their own. They learn to fish by watching older, more experienced members of their group.

In Humphrey's view, this "collegiate community" would become increasingly complex and cooperative as new environmental pressures added to the usefulness of intelligence. But this complexity brings the other side of the double-edged knife to bear upon human intelligence, and it cuts just as deep. Though the members of a social group behave cooperatively, especially toward close kin, the bottom line is that every individual is out to ensure the survival of his or her own genes. Each will be competing *within* the group for a piece of the resource pie—and that includes chances to mate, as well as opportunities to feed. Humphrey's collegiate community thus harbors the seeds of considerable political strife. As much as each individual relies on intelligence to confront the environment, his real intellectual enemies become the other members of his group. It is one thing to try to "outwit" a fig hanging just out of reach, or to fool an infant gazelle whose flight is governed by unthinking fear. But it requires a different sort of mind altogether to outwit an intelligent adversary who is simultaneously trying to outwit *you.*

Humphrey likened this intellectual challenge to the game of croquet Alice plays with the Queen of Hearts in Wonderland. Alice's mallet is a flamingo, the balls are hedgehogs, and the hoops are card men who can jump out of the way:

"You've no idea how confusing it is, all the things being alive," says poor Alice.

Without really meaning to, Nicholas Humphrey provided with his essay the harbinger of what has recently taken on the dimensions of a new paradigm for understanding the origins of our intellect, even the nature of what we call the human psyche. Some of its advocates—mostly primatologists and psychologists—call this social function of mind "Machiavellian intelligence," which should give you an idea of how much emphasis they place on the frankly political advantages of being smart. Cunning and calculation are not the only tools in the inventory of Machiavelli's model Prince, not do they exhaust the repertoire of the socially intelligent ape. Nevertheless, according to the new way of thinking, what matters is not so much *what* you know, but *whom*—and how effective you can be at keeping critical knowledge out of the hands of those who might oppose you.

Thirty years ago, when Irven DeVore, Sherwood Washburn, and others were coming back from the field with the first good data on the social behavior of baboons and other primates, the interactions between any two individuals were seen in terms of their place in a dominance hierarchy. The outcome of encounters between two individuals was fairly easy to predict: The bigger, nastier animal almost always prevailed. DeVore and Washburn's observations were not wrong, but they now seem misleadingly incomplete. Primatologists watching baboon, monkey, and chimp interactions in the wild today believe that one-on-one encounters, far from being the essence of social structure in primates, may be relatively unimportant. Rather than be doomed to submission, a physically weak member of a group might recruit stronger allies who can be depended upon to come to its aid when needed. Most often, these alliances are formed between kin.

What the primatologists are seeing in the field is basic playground politics: If somebody is bullying you, go tell your big brother so he will come stand behind you. Chances are good that the bully will gain sudden new respect for you—unless, of course, he fetches *his* brother, who happens to outrank yours in the group, further complicating the social equation. Among the primates, mothers more commonly serve as recruited allies than older siblings, but the analogy holds true. A young rhesus macaque, for instance, quickly discovers that she can intimidate members of the group far larger than herself because her high-ranking mother will always back her up in any conflict. Eventually the daughter will come to occupy a place in the female hierarchy just below her mother.

Thus the most basic kind of social intelligence is the ability to understand and remember the web of kin relationships that form the heart of a social group. What intrigues me more, however, are the political alliances that primatologists are discovering among *unrelated* individuals in a primate group, who have no shared genetic inheritance that might otherwise explain their favored treatment of each other. Primatologist Meredith Small of Cornell University recently reported the case of Becky, a low-ranking Barbary macaque in a captive community in southern France. Over a period of months, Becky managed to defy the normally rigid female hierarchy in macaques, and advanced several rungs on the social ladder by judiciously grooming and caring for the infants of females well above her in rank. To do so, Becky had to know who was powerful enough to merit the time spent currying their favor, and more importantly, she had to have some conception of a "payoff" weeks, even months down the road. Becky was one smart macaque.

In a longer-term study, Barbara Smuts of the University of Michigan found that among the fifty-odd adults in her study group in Kenya's Amboseli National Park, tight "friendships" often formed between members of opposite sexes, some lasting as long as six years. A female had one or two "special males" who tended to stay close to her even when the female was not in a mating period. When the female was attacked, her male friends rushed to her support. If the female gave birth, the male friends were often equally solicitous to her infant. In return, the special males could better count on the female's support in conflicts with strange males, and were probably favored when the time came to mate. Shirley Strum's observations of the Pumphouse Gang baboons bore this out. Generally speaking, the baboon male who mates in not the most aggressive newcomer, as Devore and Washburn believed, but the one who knows that success comes from cultivating social relationships and understanding the patterns of alliance in the group.

If the behavior of these baboons were unique, or if Becky the macaque were the only ambitious monkey forming alliances with unrelated big shots, the idea of a social origin for intelligence would not be gaining the momentum it now enjoys. But these examples just scratch the surface. All over the primate world, the reports from field biologists are giving shape and substance to Nicholas Humphrey's germ of an idea. Forming alliances is only the beginning. If it takes smarts for a baboon or monkey to keep track of all the *facts* in his social relationships, imagine how much intelligence is required when he and his companions begin to lie.

Take Paul, for instance, a young juvenile chacma baboon observed in Ethiopia by Richard Byrne and Andrew Whiten of the University of St. Andrews in Scotland. One day they noticed Paul watching an adult female named Mel dig in the ground for a large grass root. He looked around. There were no other baboons nearby, though the troop was well within earshot. Suddenly, and with no visible provocation, Paul let out a yell. In an instant his mother appeared, and in a flurry chased the astonished Mel out of sight. Meanwhile, Paul walked over and ate the grass root she left behind.

Deception is a natural art. You find it everywhere in nature: insects mimicking the appearance of plant stems or leaves, harmless snakes that look like deadly poisonous ones, fish who puff themselves up when threatened, cats who arch their backs and bristle their hair to seem bigger than they really are. On a very general level, all of these animals can be said to practice deception because they fool other animals—usually members of other species—into thinking they are something which they patently are not. Even so, it would be incorrect (not to mention odd) to accuse a stickbug or a blowfish of being a lying scoundrel. Their biology leaves them no choice but to dissemble, and so their behavior is in fact perfectly honest.

Tactical deception, to use the term coined by Byrne and Whiten, is another matter entirely. In tactical deceptions, somebody does something that fools somebody else into believing that a normal, honest state of affairs is under way, when in fact something quite different is happening. The intention to deceive is quite deliberate, and the misinterpretation is designed to play into the hands of the deceiver. Byrne and Whiten also happen to be the scientists responsible for the provocative term "Machiavellian intelligence."

"It is good to appear clement, trustworthy, humane, religious, and honest, and also to be so," Machiavelli himself advised his aspiring Prince, "but always with the mind so disposed that, when the occasion arises not to be so, you can become the opposite."

In Paul's case, the occasion that arose was his own hunger. He misled his mother into believing that he was being attacked—the "honest" reason for his scream—when in fact nothing of the kind was taking place. Again, this was not some isolated case. Paul pulled the same trick several times on other unsuspecting adults. Byrne and Whiten recently sent out a questionnaire to their primatologist colleagues, asking for similar examples of deception among primates. The responses flooded back—dozens of anecdotes telling of one chimp, baboon, or monkey duping another, or concealing some piece of information vital to his own advantage.

Sometimes it was simply a matter of one animal hiding a choice bit of food from the awareness of those in the group who were strong enough to take it away. In one case, a subordinate chimpanzee, aroused by the presence of an estrus female, covered his erect penis with his hand when a dominant male approached, and thus avoided a likely attack. Another chimp, locked in a struggle for power, hid his "fear grin" by tugging at his lips, literally wiping from his face the smile that would have betrayed his anxiety to his rival. One researcher reported watching a female hamadryas baboon slowly shuffle over toward a large rock, pretending to forage, all the time keeping an eye on the harem male in the group. After twenty minutes she ended up with her head and shoulders visible to the big male, but her hands happily engaged in the illicit activity of grooming a favorite subordinate male, who was hidden from view behind the rock.

In all of these cases, the deceivers executed a prodigious mental leap. They set upon a course of action based not only on what they *knew* of the social milieu—who is strong, who is weak, who is most likely to help—but also on what they *imagined* to be going on inside another animal's head. The female hamadryas baboon knew, for instance, that she risked attack from the harem male if she were caught grooming her favorite "other male." But she also understood that she could get away with it so long as from harem male's viewpoint, no grooming was actually taking place. By placing herself within his mental space, she had become what Nicholas Humphrey calls a "natural psychologist." This is a talent—or more properly, an adaptation—that our own species has evolved to perfection.

"The pervasiveness of deception in our everyday lives," says Richard Alexander, an evolutionary biologist at the University of Michigan, "can be glimpsed by anyone willing to reflect on how often he or she bathes, shaves, puts on deodorant, makeup, or artificial eyelashes, chooses clothes with concealing and flattering effects such as shoulder pads, dons shoes with elevated heels, pops a mint into the mouth, or enters the workplace wearing a polite smile."

More than anyone, Richard Alexander has picked up on the

On the right, "Becky," a socially intelligent Barbary Macaque, ingratiates herself with the higher-ranking female on the left by tending her infant son. MEREDITH SMALL

ideas of Nicholas Humphrey and applied them directly to a scenario of our origins. Out of all the species on earth, we are the only one that has achieved what Alexander calls "ecological dominance." We can and do live practically everywhere, under almost any ecological condition, simply by eliminating whatever threat the environment might pose to our survival. The elemental forces of natural selection—pestilence, predators, parasites, drought, climate, or what Charles Darwin called "The Hostile Forces of Nature"—no longer act upon us as they do upon other animals, and have not done so for some time. Human beings deal with predators, for example, by attacking them back—and as Alexander points out, we have done so with such thorough success in the past that we must now strive to preserve their remnants. But in the process of freeing themselves from the normal, external pressures of natural selection, our ancestors created a new force to drive human evolution forward—especially the evolution of the brain.

"The only plausible way to account for the striking departure of humans from their predecessors," Alexander says, ". . . is to assume that humans uniquely became their own principal hostile forces of nature." For better or worse, mankind invented itself.

The human mind, Alexander maintains, is the product of an evolutionary trajectory—a runaway intellect fueled by "interminable and intense conflicts of interest both within and between groups." However much a group might harbor internal conflict, the social unit whose members cooperated most effectively would have a genetic advantage over neighboring groups who were less adept at getting along for a common purpose. A "balance-of-power race" between groups would thus emerge, each forced to evolve ever finer cooperative skills simply to keep from being outcompeted—not to mention annihilated—by neighbors trapped in the same evolutionary spiral. The cliché of a "vicious cycle" can be taken quite literally here, for the net result would be a species intensely loyal to fellow group members, and viciously hostile to those on the outside. Us and them.

In Alexander's scenario, the runaway human intellect snowballs through prehistory, across the vestibule of recorded time, to eventually take up residence in the complexity of modern society, with vast communities—nation-states—engaged in an ultimate balance-of-power struggle played with nuclear weapons. This grim version of our origins leans heavily upon some recent, rather bloody revelations about the behavior of wild chimpanzees, our closest genetic relatives. After long believing that chimp society was fundamentally peaceful, in the 1970s Jane Goodall discovered to her horror that chimpanzees were capable of remarkably effective aggression against neighboring chimp groups, none the less violent for being carefully planned in advance and cooperatively perpetrated. In the most celebrated case, the males of one powerful community in the Gombe National Park study area systematically hunted down, attacked, and destroyed the males of the community next door. In three years time, the neighboring group had been utterly annihilated, its home range—and many of its females—taken over by the aggressors. Such violence, never before witnessed in the history of mammal research, was not the work of some band of simian sociopaths. In subsequent years, the victorious group was itself threatened and attacked by males from another nearby community. Similar territorial battles have been observed by researchers in other areas too.

If waging war (I see no reason to call it something else) is unique to chimpanzees and humans, then you have to wonder what traits the two species might share that would make it come about. Richard Alexander's answer is *female exogamy*. In chimpanzee society, the females of a community are the ones to leave the communities of their birth when they reach maturity. In almost all other mammal societies, it is the males who emigrate, with the females staying behind to form the social core of the group. The result among the chimps is a social structure centered around related males—all of them fathers, sons, brothers, half brothers, or close cousins.

You might expect such a tightly knit genetic fraternity to behave cooperatively, and they do. Chimps tend to forage alone or in small groups, but when one locates a tree heavily loaded with ripe fruit, he will often give forth excited pant-hoots and even drum with his palms on the resonant trunk of the tree—a din that quickly brings others to the food source. I have already mentioned their ability to hunt cooperatively. Chimpanzee males even show remarkable, if hardly perfect, mutuality when it comes to sex, sometimes lining up to copulate in succession with a single estrus female. Now we know they will also cooperate to kill their own kind when it suits their territorial interests.

"No wonder [chimpanzees] are territorial," explains field biologist Michael Ghiglieri, who has studied the chimpanzees of the Kibale Forest in Uganda. "If they were pacifists, or even individualists, their more coordinated neighbors would carve their territory into parcels and annex them. Thus armies are introduced into the natural arms race. Once this happens, solidarity between a community's males becomes essential."

Sound familiar? It should come as no great surprise that most human societies are also based on female exogamy, the males retained in the natal group to form potently cooperative, territorial units. Most modern hunter-gatherer groups also share the chimpanzee's "fusion-fission" pattern of social organization—small, impermanent groups in a community constantly separating and coming together again as resources allow. Among both chimps and humans, these reunions are often accompanied by noisy and elaborate displays of greeting, reaffirming the political unity of individuals who in fact are rarely all assembled in one place at one time.

Alexander's theory places a high premium on the social skills of manipulation and deceit. (Indeed, he believes it quite impossible that a thinking organ as powerful as the human mind would have evolved in a universe where everyone told the truth—there would simply be no need for it.) The best deceivers would be those skilled in *self*-deception, for no ruse is so successful as the one where even the deceiver is unaware of his own ulterior motives, and thus avoids the risk of betraying them. Self-deception would carry a powerful advantage in intergroup conflicts, especially as groups became increasingly larger. The members of small social units would be predisposed to cooperative behavior simply because they are closely related to each other. But as group size grew, the *us* in "us and them"—one's clansmen, countrymen, whatever—would no longer be close kin, and might in fact be total strangers. Any self-deception that would promote unity under such circumstances, such as a binding belief in tribal ritual and myth, organized religion, patriotism or ideology, would give the group a selective advantage in its conflicts with other groups. Even the human enjoyment of group-against-group forms of play—including modern sports—can be seen as a kind of no-risk practice for serious and deadly conflict.

Throughout this chapter I have warned against the dangers of drawing conclusions about human origins by thinking backward from modern humans. It would be easy to dismiss this harshly expedient view of the evolution of humanity as just another narrative told with its conclusion already in mind—Raymond Dart's killer ape again, warmed up with some sociobiological motivation and granted a scheming mind to go with his bone bludgeon. Richard Alexander admits that there is no unequivocal evidence in the fossil record revealing that our intellects rose on a tide of violence, manipulation, and deceit. But he quickly adds that there is no evidence to the contrary either. He can cite the mass of material documenting the ubiquity of warfare and aggression since the beginning of recorded history. In the modern world he can turn for support to the savage lives of the Yanomamö of the Amazon, where a quarter of all males die in warfare, or to the Hewa of New Guinea, where the murder rate is reported to be eight people per thousand per year, or for that matter to Auschwitz, Nigeria, or Cambodia. He can then point to the newly discovered belligerence of chimpanzees, and declare that the burden of proof lies with those who insist that long ago when the curtain was drawn, Lucy's Children matured under a halo of altruism and mutual trust.

At the very least, Alexander's theory, and the recent emphasis on social intelligence in general, suggest some exciting new questions for research. Did the origins of the "Machiavellian" mind have anything to do with the sudden emergence of *Homo erectus,* or were humans in the early Pleistocene still struggling more with the environment than with each other? When did our ancestors achieve "ecological dominance"? What about brain size—did it grow gradually through human evolution, or did it accelerate through time? If humans and chimpanzees exclusively share the social conditions for an intellect-enhancing balance-of-power race between groups, then what accounts for the cleverness of the other great apes—or for that matter, the keen intelligence of dolphins?

Before we can answer these questions, we need to know more

about the environments that our ancestors lived in. We need more knowledge on the intelligence of living primates, and how it is shaped by social conditions. We need to better understand the structures of the human mind, the roots of consciousness, the origins and evolution of language. We must have more information about human group sizes and population densities on the ancient landscapes. We have to comprehend more about the genetics of behavior, and the way evolution proceeds in a species struggling free of the mechanisms of natural selection that bind all others to their habitats. Above everything else, as a prerequisite to progress, we need more fossils—fossils that can tell us more about what our ancestors really looked like and how they might have behaved. Every specimen we have in hand is like a brushstroke on a canvas that has been bleached out, pulled to pieces, and scattered by the wind. We need to find more of the canvas and re-create its original colors. We need to know more. We need to know.

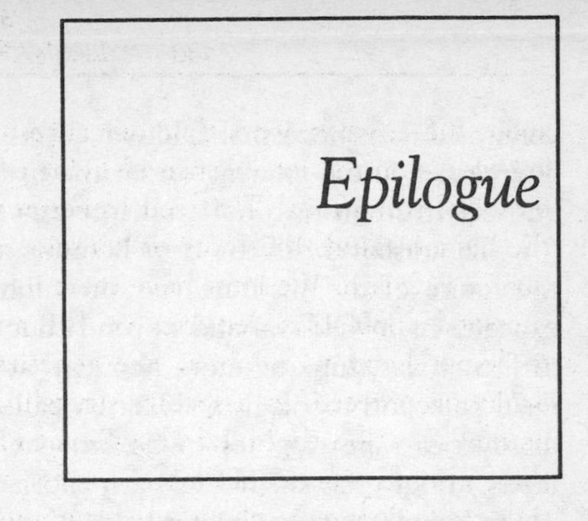

Epilogue

Of all the wonders, none is more wonderful than man,
Who has learned the art of speech, of wind-swift thought,
And of living in neighborliness.

—Sophocles, *Antigone*

THIS STORY BEGAN IN A hotel room in Dar es Salaam. Scarcely a month later I was back in Dar again, staring out a window of the Embassy Hotel. The city moved through its day with the same, familiar rhythms of amiable confusion and ragged purpose. On the crowded sidewalks, street vendors displayed frankly utilitarian objects on wobbly card tables: pads of colored paper, sewing needles, thermos liners, secondhand rubber stamps. In the streets, the overstuffed buses lay in traffic jams like beached whales, waiting for a light to change and carry them away on the flow.

All was as it had been, except for one tiny addition. Ten blocks away, a handful of blackened bone rested in a locked safe at the National Museum. I was waiting for a VW minibus to take me to the airport, with a stop at the museum to pick up the fossil. The day before, the Department of Antiquities had agreed to loan the skeleton to the Institute of Human Origins to complete research on the specimen, publish an analysis, make casts, and give American scientists a chance to examine the original. The Dik-dik Hill hominid would then return to a permanent home in the museum in Dar. The Tanzanians were delighted by the summer's success. The discovery confirmed their hope that Olduvai Gorge would have a future as well as a past, a future growing under the management of Tanzanian scientists with their own national interests at heart.

The discovery of the new hominid would mean a great deal to the future of the Institute too, in winning support for subsequent expeditions to Olduvai and beyond. But as I sat in that hotel lobby in Dar, I was thinking more about what it meant to me personally. Field paleoanthropologists are often called "fossil hunters," as if our discoveries were trophies to be brought back and held up as testimony to our prowess. If I felt any satisfaction, it was not the hubris of the hunter, but the quieter pride of someone who had participated in a successful rescue mission. We had snatched those bones from the nameless obscurity that awaited them, and given their long-vanished owner a new, empirical identity. Two millions years after her death, this little female's remains had taken on a meaning incalculable to those of her time. We had given her a part in the organized wonder we call science.

But pride was the least of my emotions. I knew that a more important rescue mission had taken place, one in which the roles were reversed. In a real sense, *I* was the one imperiled a few weeks before, and it was the Dik-dik Hill hominid who had done the rescuing. With the fossil in hand, what I felt most was *gratitude*, understandably to Tim and to the others who had worked so hard to bring it to light, but in an oddly personal way to the hominid itself, who had shown the consideration to turn up just in time to save me from my own devils. For years I had been pinning my career on a return to Ethiopia, the locus of my early success. Over time that success had become a kind of curse, as if my future were in hock to my past, with "another Lucy" the only hope of redemption. This hominid was not Lucy. She was a fascinating new piece of the human puzzle, but one that would generate as many bewildered looks as cries of "Eureka." Precisely for that reason, she was much more typical of the way this science proceeds toward its goal: each new discovery a tiny prism to catch the light from what we know and split it apart into new patterns of possibility, sharper questions, more certain uncertainties. Thanks to the Dik-dik Hill hominid, I was part of that process again. To have organized the expedition that found the skeleton, to have experienced the exuberance and the disappointments in the field, made me feel like a scientist again, not a "spokesperson." I did not need another Lucy anymore. I just needed a chance to work.

A day later, I was flying across the Atlantic, sipping a glass of wine and listening to *Tosca* over the earphones, the hominid in an orange North Face bag beneath my seat. When I got back to Berkeley, one of my first acts was to write Mary Leakey to let her know about the discovery. Late in September I received a brief note of congratulations from her.

"The last hominid allocated by me was OH61," she wrote. "Your skeleton thus becomes OH62."

So now the specimen had an official name. What it lacked in romance it more than made up for in economy: "The Dik-dik Hill hominid" was too long a label for so tiny a creature.

Through the fall Gen Suwa worked on piecing together the tooth fragments, trying to complete as much of the puzzle as was humanly possible. His efforts helped us correct a few mistakes we had made in the field when we had been joining fragments without the advantage of casts of other hominids to refer to. But nothing Gen discovered overturned our initial characterization of the fossil as a female *Homo habilis*. In May we announced the find in *Nature*. There, and in the press reports that followed, we told it the way we saw it: To judge from this specimen, in body size and build the first human species was a lot more primitive than had been thought.

We were curious about how our colleagues would receive the new addition to the hominid family. One of the first to pay it a visit was Henry McHenry, a paleoanthropologist at the University of California, Davis. As it happened, Henry had just completed an extensive study estimating the body sizes of early hominid species, including *habilis*. When I laid the bones of OH62 out on a table for him, Henry laughed.

"Such a tiny body!" he said. "It looks like I've got some more work to do." Fortunately, there was enough time before it was published for Henry to revise his study to include mention of OH62.

Meanwhile, the skeleton's long arms did not go unnoticed by the combatants in the debate over how much time our ancestors were spending in the trees. Randy Susman and Jack Stern had earlier analyzed the original *habilis* foot and hand bones from Olduvai, and had concluded that the first *Homo*, like earlier hominid species, was still committed to a partially arboreal life-style. Needless to say, they were delighted by our discovery.

"People said we were crazy," said Randy in an interview with *Science* magazine. "But now look at OH62. It's wonderful. The very long arms go perfectly with what we've said about the hands."

"*I've* looked at OH62, and I *still* think they're crazy," said Owen Lovejoy when he read that article. Owen had first seen the skeleton a few weeks after we'd returned from the field. Nothing about it made him an ounce more charitable than before to the Stony Brook team.

"They've got it ass-backwards again," he said. "Sure, OH62's arms are long relative to modern humans, just like Lucy's. But what matters is that they are *shorter* than the arms of their ape ancestors. If a species is still climbing in the trees, but it has sacrified to bipedality all the advantages of a lower limb adapted for climbing, then why on earth would it further hobble itself by reducing the length of its arms?"

Clearly, it will take more than one new fossil to settle this dispute. Perhaps Bernard Wood's evaluation of OH62 comes closest to the truth:

"The new find rudely exposes how little we know about the early evolution of *Homo,*" he wrote in *Nature*.

Since the summer of 1986 I have led two more expeditions to Olduvai. Meanwhile Prosper Ndessokia began to direct his own field team at Laetoli. One of our long-standing goals was to train local scholars to conduct their own research projects, and it was very gratifying to visit Prosper's excavations and see the *A. afarensis* fossils his team was finding. In 1988, Prosper also found a hominid lower molar near the type site of *H. habilis* in the Gorge. The same summer, high up in Olduvai Bed III deposits near the Castle, we discovered the fossil skull of a giant baboon that had been washed out by the spring rains. I had surveyed the area the year before, and I would almost swear that the baboon skull was not visible then. Soon the rains will come again to Olduvai. I wonder what treasures they will wash out next.

During those expeditions I would often pass by Dik-dik Hill. The edges of our excavation were already eroding, and the mound of backfill we had accumulated was compacted by the weather, rapidly becoming just another feature in the landscape. I could not go near the place without feeling an itch, a tickling absence. After two years, I still had an urge to find that missing distal femur of OH62. We had already announced the find and speculated about its weirdly apelike build. But without the rest of that thighbone, we were resting on hunches—educated hunches, but hunches just the same. How nice it would be, I thought, if I could sit down here on a lump of basalt, reach down, and pull that distal femur out from under a sisal bush. But OH62 had already obliged us as much as she was ever going to.

Without the rest of the bone, there was still a way we could firm up our estimate of the femur's length, which by extension would tell us how long the hominid's arms had been in proportion to her legs. The shaft of thighbone we had in hand revealed many important anatomical landmarks—holes where blood vessels passed through, for instance, and grooves, crests, and ridges where muscles had once attached. We would easily measure the distances between these various landmarks. And we could do the same on whole femurs, whose overall lengths were known. By plotting some ratios, we could work backward and predict the length of OH62's femur based on the distances between key anatomical landmarks on the bone. In essence, we could "find" that missing distal femur statistically. We would need to conduct the same kind of analysis on the humerus too. Though its shaft was intact, we lacked the articular ends.

To do the job right, we had to allow for the natural variation in bone proportions that exist within any sampling. We would have to take measurements on scores of femurs—modern humans, and chimps too—and calculate a range, which could then be applied to OH62. The best place I knew for that sort of work is the Hamann-Todd Anatomical Collection—racks upon racks of catalogued bones, housed in the basement of the Cleveland Natural History Museum, where I had started my career twenty years before. Cleveland held another lure. Owen Lovejoy could come up from Kent State and lend his considerable intellect to the analysis. Bruce Latimer, Owen's former student and curator of the Hamann-Todd Collection, could contribute his expertise as well.

In October of 1988, Tim White, Bill Kimbel, and I, accompanied by Israeli anthropologist Yoel Rak, flew out to Cleveland with the OH62 bones. Owen and Bruce met us in the museum lab.

"How do you know that all these bones belong to the same creature?" Owen challenged, even before we'd had a chance to say hello. He had hinted at this possibility before—that what we were calling OH62 was really a mix from more than one skeleton. Wasn't the radius too big for the ulna? The tibia too large to match the rest? Of course, if we really had found the remains of more than individual, then all our speculations about limb proportions were so much wind.

"Come on, Owen," I said. "What are the chances that there would be two hominids lying out there so close together? We've got no duplication of parts. The same color, the same degree of wear—it all adds up to one hominid."

"Prove it to me."

We attacked that problem first, taking measurements on dozens of chimp and human skeletons randomly selected from the collection. If the dimensions of OH62's bones, matched up against each other, fell out of the range of variation in our random sample, then we had a real problem. After the better part of two days of work, Owen was convinced.

"Okay, one hominid," he said. "But you still don't know how long the femur is."

"That's what we came here to find out," said Tim.

We got back to work. It did not take long to pin down the length of OH62's humerus to a range of only a few millimeters. The femur was trickier. After a couple of days spent hunched over our calipers, we realized that the fragmentary nature of the bone would never let us accurately reconstruct its exact length. But based on the distances calculated between a number of different anatomical landmarks, we could estimate the femur was slightly longer than Lucy's, somewhere between 300 and 330 millimeters, or roughly a foot long. When we compare the humerus of OH62 to its femur, it proved to be about 85 percent as long, suggesting proportions much like Lucy's.

Before returning to Berkeley, we had the chance to put the fossil through one more series of tests. A former student of Lovejoy's, Jim Ohman, had taken a job at Picker International, Inc., a leading manufacturer of CT scanning machines and other X-ray medical equipment. In Picker's factory just outside of Cleveland, Ohman had a demo of their latest, most powerful machine set up and ready to go. Picker had generously offered to lend us some time on the machine.

We were thrilled. With a CT scan, you can look *through* fossils at their interior structure. You could do the same with a hacksaw, of course—but no one would ever let you near a museum again. CT scans do no damage to the specimens. In the past few years, they have often proved useful in rooting out otherwise unreachable information beneath a fossil's surface. It was only on the basis of a CT scan, for instance, that the four-million year-old femur fragment that Tim had found in 1981 was proved to be that of a bipedal animal—the oldest sure hominid known. The main problem with the technique is its cost—hence our gratitude to Picker for the loan.

On a rainy morning, Bruce Latimer and I drove out to the Picker plant, a huge, hangarlike building filled with a dull hum of activity. Bruce introduced me to Jim Ohman.

"The machine we'll be using is a Picker 1200 Expert, absolutely state-of-the-art," Jim told me. "There is no other machine available with this kind of resolution."

News had gotten out that a two-million-year-old human ancestor was going to be paying a visit, and several engineers were already gathered around the demo machine when we arrived. The scanner was huge, sleek, and black—almost regally high-tech. The centerpiece was a seven-feet-tall doughnut, called the gantry, surrounding a long platform. Housed in the gantry was a laser X-ray emitter and detectors. The whole unit could move up and down the platform, while the laser spun around to project from any angle. Each beam of the X ray through the subject on the platform—in normal circumstances, a patient's head—provided a cross-section of its interior. By compiling the data from a series of these "cuts," a computer attached to the unit could generate a three-dimensional view of interior anatomy.

It was into this futuristic device, designed for medical specialists to reach into the living human brain and pinpoint abnormalities, that we placed our ancient *Homo habilis*. We started with the humerus. Jim devised a cradle to hold the specimen steady under the laser beam, and I carefully set the bone in place. He made a series of adjustments to the controls, and then another engineer turned on the laser. We watched through a protective screen as the thin red beam struck the fossil at one end, moved a millimeter forward and spat out again, proceeding down the whole length of the shaft. The procedure took all of twenty minutes; ten minutes later, we had a clear 3-D image of the fossil dancing on a video screen. By sending directions to the computer, we could turn the image around, magnify it, stand it on end, or peer into its depths at any point we wished. After we'd gleaned what data we could from the humerus, we moved on to the ulna, then to the radius and the rest of the skeleton. The cortex of the arm bones appeared to be amazingly thick—more robust even than that of many chimpanzees. OH62 may have been a little hominid, I thought, but she was far from defenseless.

In the span of a few hours we had completed the electronic metamorphosis of the fossil, her most intimate dimensions securely reiterated onto hard disk. At some point along the way I was struck by the dizzying symmetry of this event: On a gray, wet October morning in Cleveland, a sample of the most sophisticated technology

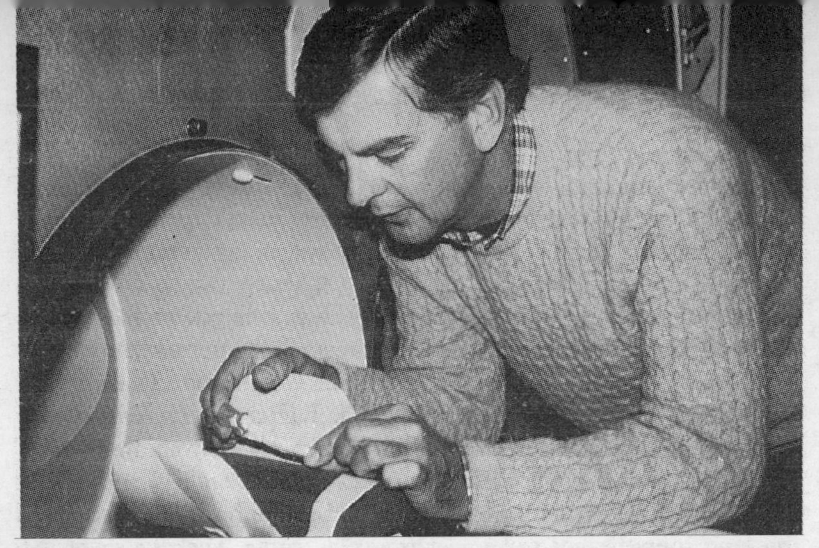

Donald Johanson prepares the ulna of OH62 for CT-scanning. BRUCE LATIMER

ever devised was whirling around this little fossil, trying to help us fathom what made the first toolmaker tick. I listened in the darkness to the gentle whirs and throbs of the equipment, and to the engineers murmuring to each other in the argot of their calling. Each time Jim Ohman placed a bone in the cradle, I thought back to when we had pulled it out of the Olduvai earth—the quick joy of discovery, and the frustration that followed. I remembered too the first morning of surveying, when I had picked up that quartz hand ax off the slope at HWK, and felt in its heft the urgency of an ancient purpose. If the Dik-dik Hill hominid were not just a collection of bones, I wondered, but instead were lying here whole and alive on the platform, would she be as impressed by *our* tools, and by what we had made of the world?

I found myself thinking about another moment later on in the field season. One afternoon, we had quit work on Dik-dik Hill a little early and decided to take a ride out to a place called Shifting Sands, about five miles west of the Olduvai camp. It had been a particularly frustrating, particularly tedious day on the site—hours spent staring at dust and pebbles, silently picking out bone chips that we knew were meaningless to our purpose. We just wanted to relax and watch the sun set over the Serengeti.

Shifting Sands is a moving mountain—or more properly speaking, an "active dune." It was formed some two thousand years ago out of a shroud of black ash that lay across the plain miles to the east—very fine, windblown volcanic sand that pulled up into a bulky mass and began to march westward before the prevailing wind. As it crawled along (just over one and a half inches a day), the dune left behind two parallel sand ridges that extend from its flanks—sand that has been blown off by crosswinds and trapped in the grass. A person looking down from a plane can trace these ridges back two miles. As the dune moves along, it continues to lose its substance to the wind. Not long ago, Shifting Sands was only one of a number of such dunes dotting the plain in this area. From the air you can see the evidence of little geological tragedies—parallel tracks of sand, tapering to a "V" that marks the spot where the other dunes ended their brief journey and vanished.

We climbed up to the top of the dune and looked to the west. The sun obliged and descended gorgeously, like a thick bronze ball melting into the edge of the earth. But to me, the sun seemed to take with it more than the light. As soon as it slipped over the horizon, I felt a sudden draining of energy, as if the tiredness in my own body were somehow connected to a falling off, a giving up, in the whole visible world around me. While the others in the group slid down the windward side of the dune, I slipped down the front, into the leeward cul-de-sac formed by its flanks. I could hear Tim White and Bob Walter talking on the other side, but the sound of their words had no resonance, no depth.

It was a place that inspired bleak thoughts. I sat there on the ground, my back up against the cold black wall of sand, and I began to wonder why I was doing what I did, what purpose this obsessive search could really serve. You find the funding, you assemble a team, you work through the politics, then you fly across the ocean, and if you're lucky you find a scrap of bone or a tooth. When you get back, you poke and probe it and make as big a deal about it as possible. A question seemed to come from the depths of the dune itself. *Why do it?* I thought about going into some other kind of work, though I could not think what it would be.

Two years later, as I sat there in the Picker International factory watching the luminous, purloined images of OH62 coalesce on a video screen, I thought I could answer the question posed by the dune. The whole human career is the story of a species never relenting in its effort to pull itself free, by the sheer force of intellect, from the natural constraints that bind all other species to their biological fates. In one sense, that is a fantastic triumph—witness the machine before me, that would soon be leaving the factory to begin its real job of saving lives—lives that might have been lost only a few years before. But the cost of this "ecological dominance" has been equally fantastic. In playing out the full extent of our uniquely unaccommodating relationship with the environment, our species has engendered a powerful self-deception: the belief that we were separately created for a special destiny, with all other life on earth subservient to our own needs.

This self-deception must once have carried with it a real evolutionary advantage, helping our species spread over the globe and overcome the challenge of virtually every known habitat. But the rules are changing. In the last few decades, our belief in the earth as *merely* a resource for humanity has led us to waste and destroy whole ecosystems, even poison the biosphere itself. The idea of a special human destiny has never been true—now it is no longer adaptive either. Simply put, we are losing our ecological dominance. The environments we have created are becoming uninhabitable, and we may not be able to control them any longer. The ozone layer above us grows thin and tears, because our life-styles depend upon refrigerators, air-conditioning and Styrofoam cups. The Greenhouse Effect may or may not be a real threat to global agriculture in the next few decades, but either way, we do not appear able to do much about it. The "us and them" balance-of-power race between human groups that Richard Alexander talks about threatens to lose its adaptive value too. No matter whose side you are on, nuclear war will hardly increase your reproductive potential.

Under these circumstances, any effort to get the truth out about who we really are and where we fit into the rhythms of the earth seems to me passionately worthwhile. I would like to think that my vocation, the search for our origins, is a part of that truth-quest. It is not simply the romantic dalliance of a species in love with itself, a species that, having led the rest of creation to the edge of annihilation, has nothing better to do than dawdle in front of a mirror, reimagining its rise to greatness. Our shared curiosity about our beginnings is the emanation of a deeper urge, felt as wonder, to discover the true humanity that lies beneath and beyond the old deceptions, the outmoded beliefs. To my mind, the primitive little human we discovered at Olduvai Gorge underscores how closely and how recently we really are bonded to our primate pasts, and by extension, to the rest of life on the planet. It is not the final word on the subject, because there are no final words. Human origins will always be enigmatic. But if we keep up the search, each new fossil we find will tell us something more. I hope we will listen to what they say.

Bibliography

Of General Interest

Johanson, D. C., and Maitland Edey. *Lucy: The Beginnings of Humankind.* New York: Simon and Schuster, 1981.

Lewin, Roger. *Bones of Contention.* New York: Simon and Schuster, 1987.

Pfeiffer, John. *The Emergence of Humankind,* 4th ed. New York: Harper and Row, 1985.

Reader, John. *Missing Links.* Boston: Little, Brown and Co., 1981.

In addition to the popular works above

Alexander, Richard. "The Evolution of the Human Psyche," in *The Human Revolution,* eds. P. Mellars and C. Stringer. Princeton, N.J.: Princeton University Press, 1989.

Ardrey, Robert. *African Genesis.* New York: Atheneum, 1961.

Binford, Lewis R. *Bones: Ancient Men and Modern Myths.* New York: Academic Press, 1981.

———. *Debating Archeology.* New York: Academic Press, 1989.

———. "Human Ancestors: Changing Views of Their Behavior," *Journal of Anthropological Archeology,* 1985, 4:292–327.

———. *In Pursuit of the Past.* London: Thames and Hudson, 1983.

Blumenschine, Robert J. "Characteristics of an Early Hominid Scavenging Niche," *Current Anthropology,* August–October 1987, 28:383–406.

Bowler, P. J. *Evolution: The History of an Idea.* Berkeley, Calif.: University of California Press, 1984.

Brain, C. K. "The Evolution of Man in Africa: Was It a Consequence of Caiozoic Cooling?" Alexander L. du Toit Memorial Lecture No. 17, Geological Society of South Africa, 1981.

Brown, Frank, John Harris, Richard Leakey, and Alan Walker. "Early *Homo erectus* skeleton from west of Lake Turkana, Kenya," *Nature*, August 29, 1985, 316:788–792.

Bunn, Henry T. "Animal Bones and Archeological Inference" (review of *Bones: Ancient Men and Modern Myths* by Lewis R. Binford), *Science*, January 29, 1982, 215:494–495.

———. "Archaeological Evidence for Meat-Eating by Plio-Pleistocene Hominids from Koobi Fora and Olduvai Gorge," *Nature*, June 18, 1981, 291:574–577.

Bunn, Henry T., and Ellen M. Kroll. "Systematic Butchery by Plio-Pleistocene Hominids at Olduvai Gorge, Tanzania," *Current Anthropology*, December 1986, 27:431–452.

Byrne, Richard, and Andrew Whiten, eds. *Machiavellian Intelligence*. Oxford, England: Oxford University Press, 1988.

———. "The Thinking Primate's Guide to Deception," *New Scientist*, December 3, 1987, 54–57.

Cheney, D., R. Seyfarth, and B. Smuts. "Social Relationships and Social Cognition in Nonhuman Primates," *Science*, December 12, 1986 234:1361–1366.

Coles, Sonia. *Leakey's Luck*. New York: Harcourt Brace Jovanovich, 1975.

Dart, Raymond. *Adventures with the Missing Link*. Philadelphia: The Institutes Press, 1967.

Day, M. H., M. D. Leakey, and T. Olson. "On the Status of *Autralopithecus afarensis*," *Science*, March 1980, 207:1102–1103.

Delson, Eric, ed. *Ancestors: The Hard Evidence*. New York: Alan R. Liss, Inc., 1984.

Foley, Robert. *Another Unique Species*. Essex, England: Longman Press, 1987.

Foley, Robert, ed. *Hominid Evolution and Community Ecology*. London: Academic Press, 1984.

Ghiglieri, Michael. "The Social Ecology of the Chimpanzees," *Scientific American*, 1985, 252:102–113.

———. "Sociobiology of the Great Apes and the Hominid Ancestor," *Human Evolution*, 1987, 16:319–357.

Goodall, Jane. *The Chimpanzees of Gombe*. Cambridge, Mass.: Harvard University Press, 1986.

Harcourt, Alexander. "All's Fair in Play and Politics," *New Scientist*, December 12, 1985.

Harding, R.S.O., and G. Teleki, eds. *Omnivorous Primates*. New York: Columbia University Press, 1981.

Hay, Richard. *Geology of Olduvai Gorge*. Berkeley, Calif.: University of California Press, 1976.

Hay, Richard L., and Mary D. Leakey. "The Fossil Footprints of Laetoli," *Scientific American*, February 1982.

Herbert, Wray. "Lucy's Family Problems," *Science News*, July 2, 1983, 124:8–11.

Howell, F. Clark. *Early Man*, rev. ed. New York: Time-Life Books, 1973.

Isaac, Glynn. "The Food-Sharing Behavior of Protohominids," *Scientific American*, April 1976, 90–108.

Isaac, Glynn, and Elizabeth R. McGown, eds. *Human Origins*. Menlo Park, Calif.: W. A. Benjamin, 1976.

Johanson, D. C., and M. Taieb. "Plio-Pleistocene Hominid Discoveries in Hadar, Ethiopia," *Nature*, March 1976, 260:293–297.

Johanson, Donald C., Fidelis T. Masao, Gerald G. Eck, Tim White, Robert C. Walter, William H. Kimbel, Berhane Asfaw, Paul Manega, Prosper Ndessokia, and Gen Suwa. "New Partial Skeleton of *Homo habilis* from Olduvai Gorge, Tanzania," *Nature*, May 21, 1987, 327:205–209.

Johanson, D. C., and T. D. White. "On the Status of *Australopithecus afarensis*," *Science*, March 1980, 207:1104–1105.

———. "A Systematic Assessment of Early African Hominids," *Science*, January 1979, 203:321–330.

Johanson, Donald C., et al. "Pliocene Hominids Fossils from Hadar, Ethiopia," *American Journal of Physical Anthropology*, April 1982, Vol. 57, No. 4 (complete issue).

Keeley, Lawrence K., and Nicholas Toth. "Microwear Polishes on Early Stone Tools from Koobi Fora, Kenya," *Nature*, October 1981, 293:464–465.

Kimbel, W. H., and T. D. White. "A Reconstruction of the Adult Cranium of *Australopithecus afarensis*," *American Journal of Physical Anthropology*, February 1980, 52:244.

Kimbel, W. H., T. D. White, and D. C. Johanson. "Implications of KNM-WT 17000 for the Evolution of 'Robust' *Australopithecus*," in *The Evolutionary History of the "Robust" Australopithecines*, ed. F. E. Grine, in press.

Kinzey, W., ed. *Primate Models of Hominid Behavior*. New York: Plenum Press, 1985.

Leakey, L.S.B. "Exploring 1,750,000 Years into Man's Past," *National Geographic*, October 1961, 120/4:564–589.

———. "Finding the World's Earliest Man," *National Geographic*, September 1960, 118/3:420–435.

Leakey, L.S.B., P. V. Tobias, and J. R. Napier. "A New Species of the Genus *Homo* from Olduvai Gorge," *Nature*, April 1964, 202:7–9.

Leakey, M. D. *Olduvai Gorge, Volume 3*. Cambridge, England: Cambridge University Press, 1972.

Leakey, M. D., R. L. Hay, G. H. Curtis, R. E. Drake, M. K. Jackes, and

T. D. White. "Fossil Hominids from the Laetoli Beds," *Nature*, August 1976, 262:460–466.

Leakey, Richard E. "Hominids in Africa," *American Scientist*, March–April 1976, 64:174–178.

Leakey, R.E.F., and A. Walker "On the Status of *Australopithecus afarensis*," *Science*, March 1980, 207:1103.

Lewin, Roger. "The Origin of the Human Mind," *Science*, May 8, 1987, 236:668–669.

Lovejoy, C. Owen. "Evolution of Human Walking," *Scientific American*, November 1988, 118–125.

———. "The Gait of *Australopithecus*," *American Journal of Physicial Anthropology*, 38:757–780.

Olson, T. "Basicranial Morphology of the Extant Hominoids and Pliocene Hominids," in *Aspects of Human Evolution*, ed. C. Stringer. London: Taylor & Francis, 99–128.

Potts, Richard. "Home Bases and Early Hominids," *American Scientist*, 1984, 72:338–347.

Potts, Richard, and Pat Shipman. "Cutmarks Made by Stone Tools on Bones from Olduvai Gorge, Tanzania," *Nature*, June 18, 1981, 291:577–580.

Rak, Yoel. "Australopithecine Taxonomy and Phylogeny in Light of Facial Morphology," *American Journal of Physical Anthropology*, 1985, 66:281–287.

———. "Lucy's Pelvic Anatomy: Its Role in Bipedal Gait," unpublished manuscript.

Shipman, Pat. "The Ancestor That Wasn't," *The Sciences*, March–April 1985, 43–48.

———. "Baffling Limb on the Family Tree," *Discover*, September 1986.

———. "Scavenger Hunt," *Natural History*, April 1984, 20–27.

Shipman, Pat, Wendy Bosler, and Karen Lee Davis. "Butchering of Giant Geladas at an Acheulian Site," *Current Anthropology*, June 1981, 22/3:257–264.

Small, Meredith. "Ms. Monkey," *Natural History*, January 1989.

———. "Social Climber: Independent Rise in Rank by a Female Barbary Macaque," unpublished manuscript.

Stern, Jack T., and Randall L. Susman. "The Locomotor Anatomy of *Australopithecus afarensis*, *American Journal of Physical Anthropology*, 1983, 60:279–317.

Stringer, C. B. "The Credibility of *Homo habilis*," in *Major Topics in Primate and Human Evolution*, eds. B. Wood, L. Martin, and P. Andrews. Cambridge, England: Cambridge University Press, 1984.

Susman, Randall L., Jack T. Stern, and William L. Jungers. "Arboreality and Bipedality in the Hadar Hominids," *Folia primatologica*, 1984, 43:113–156.

Tattersall, Ian. "Species Recognition in Human Paleontology," *Journal of Human Evolution,* 1986, 15:165–175.

Vrba, Elisabeth. "The Environmental Context of the Evolution of Early Hominids and Their Culture," in press.

———. "Late Pliocene Climatic Events and Hominid Evolution," in *The Evolutionary History of the "Robust" Australopithecines,* ed. F. E. Grine, in press.

Walker, Alan. "Extinction in Hominid Evolution," in *Extinctions,* ed. M. H. Nitecki. Chicago: Chicago University Press, 1984, pp. 119–152.

Walker, A., R. E. Leakey, J. M. Harris, and F. H. Brown. "2.5-Myr *Australopithecus boisei* from west of Lake Turkana, Kenya," *Nature,* Augsut 7, 1986, 322:517–522.

White, T. D., and J. Harris. "Suid Evolution and Correlation of African Hominid Localities," *Science,* 1977, 198:13–21.

White, Tim D., Donald C. Johanson, and Wiliam H. Kimbel. "*Australopithecus africanus:* Its Phyletic Position Reconsidered," *South African Journal of Science,* October 1981, 77:445–470.

White, Tim D., and Gen Suwa. "Hominid Footprints at Laetoli: Facts and Interpretations," *American Journal of Physical Anthropology,* April 1987, 72:485–514.

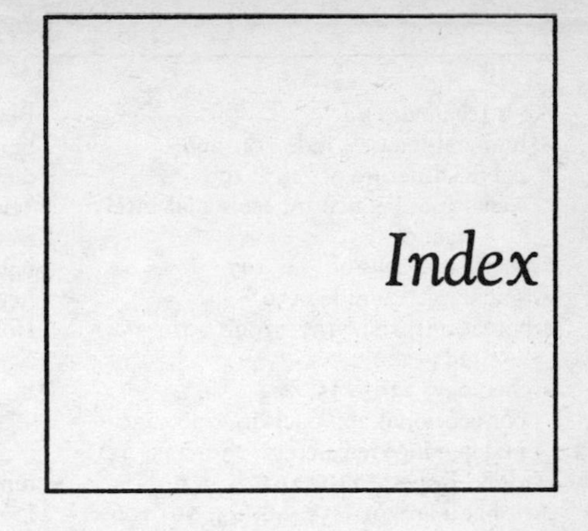

A

C

D

G

I

J

K

M

N

O

R

S

U

V

W

Y

Z

BY THE SAME AUTHOR

Lucy: The Beginnings of Humankind
by Donald C. Johanson and Maitland A. Edey

When Donald Johanson uncovered 'Lucy' in 1974 – the oldest, most complete and best-preserved skeleton of an erect-walking human ancestor we have – it caused a sensation and heralded a major scientific breakthrough. In this gripping book Johanson and Maitland Edey explain in graphic detail what was already known, how Lucy was found, what exactly she *is*, where she fits into the jigsaw of evidence, and what (despite continuing controversy) she has taught us about human origins.

'Written with all the verve and enthusiasm that the young anthropologists brought to her discovery – a discovery that was made against a background of political intrigue that sometimes reads more like a thriller' – Theya Molleston in *The Times*

'A fascinating story . . . Lucy herself is stunning' – Desmond Morris

'The best written book on anthropology I have read . . . I could not stop reading' – Adrian Berry in the *Sunday Telegraph*